Particulate Morphology

Particulate Morphology
Mathematics Applied to Particle Assemblies

Keishi Gotoh

Emeritus Professor
Toyohashi University of Technology
Toyohashi, Japan

AMSTERDAM • BOSTON • HEIDELBERG • LONDON • NEW YORK • OXFORD
ELSEVIER   PARIS • SAN DIEGO • SAN FRANCISCO • SINGAPORE • SYDNEY • TOKYO

Elsevier
32 Jamestown Road, London NW1 7BY
225 Wyman Street, Waltham, MA 02451, USA

First edition 2012

Notices
Knowledge and best practice in this field are constantly changing. As new research and experience broaden our understanding, changes in research methods, professional practices, or medical treatment may become necessary.

Practitioners and researchers must always rely on their own experience and knowledge in evaluating and using any information, methods, compounds, or experiments described herein. In using such information or methods they should be mindful of their own safety and the safety of others, including parties for whom they have a professional responsibility.

To the fullest extent of the law, neither the Publisher nor the authors, contributors, or editors, assume any liability for any injury and/or damage to persons or property as a matter of products liability, negligence or otherwise, or from any use or operation of any methods, products, instructions, or ideas contained in the material herein.

British Library Cataloguing-in-Publication Data
A catalogue record for this book is available from the British Library

Library of Congress Cataloging-in-Publication Data
A catalog record for this book is available from the Library of Congress

ISBN: 978-0-323-28257-4

For information on all Elsevier publications
visit our website at elsevierdirect.com

This book has been manufactured using Print On Demand technology. Each copy is produced to order and is limited to black ink. The online version of this book will show color figures where appropriate.

Working together to grow
libraries in developing countries
www.elsevier.com | www.bookaid.org | www.sabre.org

ELSEVIER   BOOK AID
             International   Sabre Foundation

Contents

Preface

In the field of science and technology, mathematics can be considered as a general tool for expressing a state and/or phenomena of various substances so that it is applicable to ubiquitous subjects. Accordingly, it is recommended to study as much as possible before we engage in any research subject. However, there are too many tools to learn so that we usually start our research without sufficient knowledge to deal with, except the person who has interest in mathematics itself. There are many regrettable cases in which more effective results might be derived by use of sufficient mathematical knowledge.

The author has been involved for a long time in the research of "Flow and Measurement of Particulate Materials," especially in the statistical geometry of "Spatial Structure of Random Particle Assemblies." This text is a collection of mathematical tools used in the series of papers. In other words, the content is applied mathematics concerning personal research subjects. The author will be grateful if it helps many students and young researchers make further advancement.

Mathematical derivations are shown in as much detail as possible without omission so that readers can follow them easily. References cited are listed at the end of the chapters.

The author acknowledges sincerely all staffs of Elsevier Insights for their laborious editing work.

1 Spatial Structure of Random Dispersion of Equal Spheres in One Dimension

Although the one-dimensional structure of random dispersion of equal spheres may not exist in so many in practical applications, it is useful such that we can derive the exact analytical solution and examine the accuracy of computer experiments. Also it is helpful for studying the comparison between a discrete system and a continuous one.

1.1 Discrete System

Consider a line of equal spheres placed at unit interval as depicted in Figure 1.1. Two neighboring spheres are connected one at a time by placing a stick on the interval under the condition that overlapping connection is forbidden.

We call the connected two spheres the cluster of size 2, the connected three spheres the cluster of size 3, and so on. On the other hand, we call n series of isolated spheres the isolated group of size n.

Figure 1.1 shows an isolated group of size 4 where three different cases are inherently possible to exist for the isolated group of size 2, and two different cases for the isolated group of size 3. In general, the isolated group of size n inherently contains two possible cases for the isolated group of size $(n - 1)$, three possible cases for the isolated group of size $(n - 2)$, and so on.

Consider M lines of N series of isolated equal spheres. Two neighboring spheres are connected by placing a stick on the interval repeatedly under the condition that duplicate connection is forbidden. The probability of producing a cluster of size 2 during the time interval $t - (t + dt)$ is expressed by $k(t)dt$, and the number of the isolated groups of size n at time t in the jth line by $C^{(j)}(n,t)$. Accordingly, we obtain (Cohen and Reiss, 1963)

$$-dC^{(j)}(n, t)/dt = k(t)[(n - 1)C^{(j)}(n, t) + 2C^{(j)}(n + 1, t)] \tag{1.1}$$

Particulate Morphology. DOI: 10.1016/B978-0-12-396974-3.00001-1

Size 4   **Figure 1.1** Isolated group of size 4.

Equation (1.1) expresses the decreasing rate of the number of the isolated groups of size n at time t. There are $(n-1)$ possible cases for connecting two neighboring spheres in the isolated group of size n. Therefore, the first term of the right-hand side of Eq. (1.1) expresses the total number of the possible cases that the isolated group of size n is internally destroyed by placing the stick. When either end-sphere of the isolated group of size n is connected to its outside neighbor, the group is destroyed to become the size $(n-1)$. The connection to the outside neighbor means the existence of $C^{(j)}(n+1,t)$. The other side of the isolated group is also taken into account, leading to the second term of the right-hand side of Eq. (1.1).

For M lines of N series of spheres, we define the average number of the isolated groups of size n as follows:

$$\langle C(n,t)\rangle = M^{-1}\sum_{j=1}^{M} C^{(j)}(n,t) \tag{1.2}$$

Hence, Eq. (1.1) becomes

$$-d\langle C(n,t)\rangle/dt = k(t)[(n-1)\langle C(n,t)\rangle + 2\langle C(n+1,t)\rangle] \tag{1.3}$$

Because there are $(N-n+1)$ possible cases for making the isolated group of size n from N series of spheres, the initial condition becomes as follows:

$$\langle C(n,0)\rangle = N - n + 1 \tag{1.4}$$

By use of $z = \int_0^t k(t)dt$ or $dz = k(t)dt$, Eq. (1.3) is simplified to become

$$-d\langle C(n, z)\rangle/dz = (n - 1)\langle C(n, z)\rangle + 2\langle C(n + 1, z)\rangle \tag{1.5}$$

The solution of Eq. (1.5) with the initial condition (1.4) becomes

$$\langle C(n, z)\rangle = \exp[-(n - 1)z]\sum_{s=0}^{N-n}(N - n - s + 1)(2e^{-z} - 2)^s/s! \tag{1.6}$$

or

$$\langle C(n + 1, z)\rangle = \exp[-nz]\sum_{s=0}^{N-n-1}(N - n - s)(2e^{-z} - 2)^s/s!$$

Substitution of Eq. (1.6) into Eq. (1.5) can verify the solution, where replacing $(s - 1)$ by k, the summation for the derivative term should be from $k = 0$ to $(N - n - 1)$.

Accordingly, the survival rate of the isolated groups of size n becomes as follows:

$$\begin{aligned}P(n, z) &= \langle C(n, z)\rangle/(N - n + 1)\\ &= e^{-(n-1)z}\sum_{s=0}^{N-n}[1 - s/(N - n + 1)](2e^{-z} - 2)^s/s!\\ &\to e^{-(n-1)z}\sum_{s=0}^{N-n}(2e^{-z} - 2)^s/s! : \quad N \to \infty\end{aligned} \tag{1.7}$$

$$= e^{-(n-1)z}\exp[-2(1 - e^{-z})] \tag{1.8}$$

The survival probability of a single isolated sphere after a long time is obtainable by setting $n = 1$ and $z = \infty$ in Eq. (1.8).

$$P(1, \infty) = e^{-2} \tag{1.9}$$

a. Case of placing the cluster of size k smaller than the isolated group $(k \leqq n)$

In the previous discussion, we considered the case of placing a single stick to connect two neighboring spheres every time. Here we consider the case of placing the cluster of size k every time under the condition that overlapping connection is forbidden, where $k \leqq n$. The survival probability of the isolated group of size n is expressed by $P(n,t)$ at time t. Then, similarly to Eqs. (1.1)–(1.3), the following relation is derived for the decreasing rate of the isolated cluster of size n (Rodgers, 1992).

$$-dP(n,t)/dt = (n - k + 1)P(n,t) + 2\sum_{j=1}^{k-1} P(n+j,t) \qquad (1.10)$$

where $k \leq n$.

In the right-hand side of Eq. (1.10), $(n - k + 1)$ expresses the possible number of cases in which the isolated group of size n is destroyed by placing the cluster of size k inside the group. The isolated group of size n is also destroyed by placing the cluster of size k partially inside the group: one side of the cluster laps by the interval j from the end point n of the isolated group. In order to make such an occasion possible, the isolated group of size $(n + j)$ does exist first and the cluster of size k is placed, where $j = 1, 2, \ldots, (k - 1)$. The second term of the right-hand side of Eq. (1.10) is for this type of the partial destruction of the isolated group of size n. Assuming

$$P(n,t) = F(t)\exp[-(n - k + 1)] \qquad (1.11a)$$

and substituting into Eq. (1.10),

$$\begin{aligned}
- dP(n,t)/dt &= -\{dF(t)/dt\}\exp[-(n - k + 1)t] \\
&\quad + F(t)(n - k + 1)\exp[-(n - k + 1)t] \\
&= (n - k + 1)F(t)\exp[-(n - k + 1)t] \\
&\quad + 2\sum_{j=1}^{k-1} F(t)\exp[-(n + j - k + 1)t]
\end{aligned}$$

$$d\ln F(t)/dt = -2\sum_{j=1}^{k-1} \exp[-jt]$$

Using the initial conditions $P(n,0) = 1$ and $F(0) = 1$,

$$F(t) = \exp\left[-2\sum_{j=1}^{k-1}(1 - e^{-jt})/j\right] : k \geq 2 \qquad (1.12)$$

Therefore, Eq. (1.11a) becomes as follows:

$$P(n,t) = \exp[-(n - k + 1)t]\exp\left[-2\sum_{j=1}^{k-1}(1 - e^{-jt})/j\right] : n \geq k \geq 2 \qquad (1.11b)$$

b. Case of placing the cluster of size k larger than the isolated group $(k \geqq n)$

In the case of $k \geqq n$, Eq. (1.10) becomes as follows:

$$-dP(n,t)/dt = (k - n + 1)P(k,t) + 2\sum_{j=1}^{n-1} P(k+j,t) \qquad (1.13a)$$

In order to place the cluster larger than the isolated group of size n without any overlap, the isolated group of size k must exist first.

There are $(k - n + 1)$ possible cases for the isolated group of size n to exist in the isolated group of size k, leading to the first term of the right-hand side of Eq. (1.13a). The second term expresses the partial destruction of the isolated group of size n by placing the cluster of size k. Consider that the cluster of size k is placed on the isolated group leaving j isolated spheres: $j = 1, 2,\ldots, (n - 1)$. The isolated group of size $(k+j)$ must exist first for this purpose. The partial destruction of the isolated group of size n is possible to occur from both ends, leading to the second term of the right-hand side of Eq. (1.13a).

In Eq. (1.11a) with Eq. (1.12), which is the solution of Eq. (1.10), the substitutions of k into n and $(k+j)$ into n give

$$P(k,t) = F(t)\exp[-(k - k + 1)t] = F(t)e^{-t}$$

$$P(k+j,t) = F(t)\exp[-(k+j-k+1)t] = F(t)e^{-(j+1)t}$$

$$F(t) = \exp\left[-2\sum_{j=1}^{k-1}(1 - e^{-jt})/j\right]$$

Hence, Eq. (1.13a) becomes as follows:

$$-dP(n,t)/dt = (k - n + 1)P(k,t) + 2\sum_{j=1}^{n-1}P(k+j,t)$$

$$= \left\{(k - n + 1) + 2\sum_{j=1}^{n-1}e^{-jt}\right\}\exp\left[-t - 2\sum_{j=1}^{k-1}(1 - e^{-jt})/j\right]$$

$$\qquad (1.13b)$$

Therefore, we obtain (Rodgers, 1992)

$$P(n,t) = 1 - \int_0^t \left\{ (k-n+1) + 2\sum_{j=1}^{n-1} e^{-jt} \right\} \exp\left[-t - 2\sum_{j=1}^{k-1}(1-e^{-jt})/j \right] dt : k \geqq n$$

(1.14a)

where the initial condition is $P(n,0) = 1$.
Especially in the case of $n = k$,

$$P(k,t) = 1 - \int_0^t \left\{ 1 + 2\sum_{j=1}^{k-1} e^{-jt} \right\} \exp\left[-t - 2\sum_{j=1}^{k-1}(1-e^{-jt})/j \right] dt$$

$$= 1 + \int_0^t \left\{ -1 - 2\sum_{j=1}^{k-1} e^{-jt} \right\} \exp\left[-t - 2\sum_{j=1}^{k-1}(1-e^{-jt})/j \right] dt$$

$$= 1 + \int_0^t \frac{d}{dt} \left\{ \exp\left[-t - 2\sum_{j=1}^{k-1}(1-e^{-jt})/j \right] \right\} dt$$

$$= \exp\left[-t - 2\sum_{j=1}^{k-1}(1-e^{-jt})/j \right]$$

From Eq. (1.14a),

$$P(1,t) = 1 - k\int_0^t \exp\left[-t - 2\sum_{j=1}^{k-1}(1-e^{-jt})/j \right] dt$$

(1.14b)

From Eq. (1.14b), for $k = 1$:

$$P(1,t) = 1 - \int_0^t e^{-t}\, dt$$

$$y(t) = 1 - P(1,t) = \int_0^t e^{-t}\, dt = 1 - e^{-t}$$

(1.15)

and for $k = 2$:

$$P(1,t) = 1 - 2\int_0^t \exp[-t - 2(1-e^{-t})]dt$$

$$y(t) = 1 - P(1, t) = 2 \int_0^t \exp[-t - 2(1 - e^{-t})] dt$$

$$= - \int_0^t d/dt\{\exp[-t - 2(1 - e^{-t})]\} dt \qquad (1.16)$$

$$= 1 - \exp[-2(1 - e^{-t})]$$

The fraction of the line covered by the cluster at time t, $y(t)$, depends on the size of the cluster, k.

The discrete system hitherto mentioned will be changed into the continuous system in the next section.

1.2 Continuous System

As depicted in Figure 1.2, equal particles of size a are randomly placed one by one on the line without any overlap. The fractional length, $y(t)$, covered with the particles at time t is considered. The particle size a is set equal to k this time, while in the previous section the size $(k - 1)$ of the cluster was equal to the particle size. Accordingly, we can use the previous results by replacing k by $(k + 1)$ and n by $(n + 1)$ in the corresponding equations.

Referring to the figure of explanation, the equations are modified to use for the continuous system.

a. In the case of $n \geqq k$, Eq. (1.11b) is made continuous as follows:

$$P(n, t) = \exp\left[-(n - k + 1)t - 2\sum_{j=1}^{k-1}(1 - e^{-jt})/j\right] \qquad (1.11b)$$

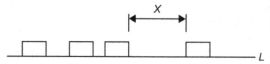

Figure 1.2 Random arrangement of particles with length a.

Replacing k by $(k + 1)$ and n by $(n + 1)$ in the right-hand side of the above equation, we obtain

$$P(n,t) = \exp\left[-(n-k+1)t - 2\sum_{j=1}^{k}(1-e^{-jt})/j\right]$$

$$= \exp\left[-(n-a+1)t - 2\sum_{j=1}^{a}(1-e^{-jt})/j\right]$$

In the continuous system, the outer unit interval [0,1] of the isolated group n belongs to the isolated group itself and therefore the size of the vacancy should be $(m + 1)$ in place of n, where m is used as the continuous variable. Namely,

$$P(m,t) = \exp\left[-\{(m+1)-a+1\}t - 2\sum_{j=1}^{a}\{(1-e^{-jt})/(jt)\}d(jt)\right]$$

$$= \exp\left[-(m-a)t - 2\sum_{j=0}^{a}\{(1-e^{-jt})/(jt)\}d(jt)\right]$$

$$P(x,t) = \exp\left[-(x-1)at - 2\int_{0}^{at}\{(1-e^{-v})/v\}dv\right] : \quad x = m/a$$

Replacing at by τ, we obtain

$$P(x,\tau) = \exp\left[-(x-1)\tau - 2\int_{0}^{\tau}\{(1-e^{-v})/v\}dv\right]$$

Hence the probability that no particle exists in the interval $[x,(x + dx)]$ becomes as follows:

$$f(x,\tau)dx = -[dP(x,\tau)/dx]dx$$

$$= \tau \, dx \, \exp\left[-(x-1)\tau - 2\int_{0}^{\tau}\{(1-e^{-v})/v\}dv\right] : \quad x \geqq 1 \qquad (1.17a)$$

Therefore,

$$
\begin{aligned}
\int_1^\infty f(x,\tau)\,dx &= \int_1^\infty \tau \, \exp\left[-(x-1)\tau - 2\int_0^\tau \{(1-e^{-v})/v\}dv\right]dx \\
&= -\int_1^\infty d/dx\left\{\exp\left[-(x-1)\tau - 2\int_0^\tau \{(1-e^{-v})/v\}dv\right]\right\}dx \\
&= \exp\left[-2\int_0^\tau \{(1-e^{-v})/v\}dv\right]
\end{aligned}
$$

(1.17b)

b. In the case of $n \leq k$, Eq. (1.13a) is made continuous as follows:

$$
-dP(n,t)/dt = (k-n+1)P(k,t) + 2\sum_{j=1}^{n-1} P(k+j,t) \tag{1.13a}
$$

In the right-hand side of the above equation, replacement of k by $(k+1)$ and n by $(n+1)$ gives

$$
-dP(n,t)/dt = (k-n+1)P(k+1,t) + 2\sum_{j=1}^{n} P(k+1+j,t) \tag{1.13c}
$$

Similar replacements for n and k in Eqs. (1.11a) and (1.12) give

$$
P(n,t) = \exp[-(n-k+1)t]\exp\left[-2\sum_{j=1}^{k}(1-e^{-jt})/j\right]
$$

Therefore,

$$
P(k+1,t) = e^{-2t}\exp\left[-2\sum_{j=1}^{k}(1-e^{-jt})/j\right] = \exp\left[-2\sum_{j=0}^{k}(1-e^{-jt})/j\right]
$$

$$
P(k+1+j,t) = e^{-jt}P(k+1,t)
$$

Hence, Eq. (1.13c) gives

$$P(n,t) = 1 - \int_0^t \left\{ (k-n+1) + 2\sum_{j=1}^{n} e^{-jt} \right\} \exp\left[-2\sum_{j=0}^{k}(1-e^{-jt})/j \right] dt$$

$$= 1 - \int_0^t \left\{ (a-n+1) + 2\sum_{j=1}^{n} e^{-jt} \right\} \exp\left[-2\sum_{j=0}^{a}(1-e^{-jt})/j \right] dt$$

The interval [0,1] belongs to the isolated group n in the continuous system and yet the group should be included in the size $k(=a)$.

Accordingly, in the first term $(a-n+1)$ of { }, n should be replaced by $(m-1)$. The second term remains unchanged and we rewrite n as the new symbol m.

$$P(m,t) = 1 - \int_0^t \left\{ (a-m+2) + 2\sum_{j=1}^{m} e^{-jt} \right\} \exp\left[-2\sum_{j=0}^{a}(1-e^{-jt})/j \right] dt$$

$$= 1 - \int_0^t \left\{ (a-m) + 2\sum_{j=0}^{m} e^{-jt} \right\} \exp\left[-2\sum_{j=0}^{a}(1-e^{-jt})/j \right] dt$$

Replacing $m/a = x$ and $at = \tau$, we make it continuous to become

$$P(x,t) = 1 - \int_0^t \left\{ (1-x) + 2\int_0^x e^{-atz}dz \right\} \exp\left[-2\int_0^{at}\{(1-e^{-v})/v\}dv \right] a\,dt$$

$$= 1 - \int_0^t \left\{ 1-x + 2(1-e^{-atx})/(at) \right\} \exp\left[-2\int_0^{at}\{(1-e^{-v})/v\}dv \right] a\,dt$$

$$P(x,\tau) = 1 - \int_0^{\tau} \{ 1-x + 2(1-e^{-\tau x})/\tau \} \exp\left[-2\int_0^{\tau}\{(1-e^{-v})/v\}dv \right] d\tau$$

$$f(x,\tau) = -[dP(x,\tau)/dx]$$

$$= \int_0^{\tau} \{ -1 + 2e^{-\tau x} \} \exp\left[-2\int_0^{\tau}\{(1-e^{-v})/v\}dv \right] d\tau \tag{1.18a}$$

$$\int_0^1 f(x, \tau)dx$$

$$= \int_0^\tau \{-1 + 2(1 - e^{-\tau})/\tau\} \exp\left[-2\int_0^\tau \{(1 - e^{-v})/v\}dv\right]d\tau$$

$$= -\int_0^\tau \exp\left[-2\int_0^\tau \{(1 - e^{-v})/v\}dv\right]d\tau$$

$$\qquad - \int_0^\tau d/d\tau \left\{\exp\left[-2\int_0^\tau \{(1 - e^{-v})/v\}dv\right]\right\}d\tau \qquad (1.18b)$$

$$= -\int_0^\tau \exp\left[-2\int_0^\tau \{(1 - e^{-v})/v\}dv\right]d\tau$$

$$\qquad - \exp\left[-2\int_0^\tau \{(1 - e^{-v})/v\}dv\right] + 1$$

Hence, we obtain the coverage fraction $y(\tau)$, i.e., the probability that the line depicted in Figure 1.2 is covered by the particles at time τ as follows (Gonzalez et al., 1974; Feder, 1980):

$$y(\tau) = 1 - \int_0^\infty f(x, \tau)dx = 1 - \left[\int_0^1 f(x, \tau)dx + \int_1^\infty f(x, \tau)dx\right]$$

Substitution of Eqs. (1.17b) and (1.18b) yields

$$y(\tau) = \int_0^\tau \exp\left[-2\int_0^\tau \{(1 - e^{-v})/v\}dv\right]d\tau \qquad (1.19a)$$

After a long time, i.e., $\tau \to \infty$, Eq. (1.19a) can be approximated as follows (Hinrichsen et al., 1986):

$$y(\tau) = \int_0^\tau \exp\left[-2\int_0^\tau \{(1 - e^{-v})/v\}dv\right]d\tau$$

$$= y(\infty) - \int_\tau^\infty \exp\left[-2\int_0^\tau \{(1 - e^{-v})/v\}dv\right]d\tau$$

where

$$e^{-v} = \sum_{n=0}^{\infty} (-1)^n v^n/n!; \quad 1 - e^{-v} = -\sum_{n=1}^{\infty} (-1)^n v^n/n!$$

$$\int_0^\tau \{(1 - e^{-v})/v\}\mathrm{d}v = -\sum_{n=1}^{\infty} \{(-1)^n/n!\} \int_0^\tau v^{n-1}\mathrm{d}v$$

$$= -\sum_{n=1}^{\infty} (-1)^n \tau^n/(n!n)$$

Bessel formula

$$\mathrm{Ei}(x) = \int_x^\infty (e^{-t}/t)\mathrm{d}t = \gamma + \ln x + \sum_{n=1}^{\infty} (-1)^n x^n/(n!n)$$

$$= \gamma + \ln x - e^{-x} \sum_{n=1}^{\infty} (x^n/n!) \left\{ \sum_{r=1}^{n} (1/r) \right\}$$

$$: \mathrm{Ei}(x) = +\infty \sim 0 \quad \text{for } x = 0 \sim \infty \quad (\text{cf. Numerical chart})$$

Introducing the above Bessel formula, we obtain

$$\int_0^\tau \{(1 - e^{-v})/v\}\mathrm{d}v = \gamma + \ln \tau - \mathrm{Ei}(\tau) \tag{1.19b}$$

$$y(\tau) = \int_0^\tau \exp\left[-2\int_0^\tau \{(1 - e^{-v})/v\}\mathrm{d}v\right]\mathrm{d}\tau$$

$$= y(\infty) - \int_\tau^\infty \exp\left[-2\int_0^\tau \{(1 - e^{-v})/v\}\mathrm{d}v\right]\mathrm{d}\tau$$

$$\to y(\infty) - \int_\tau^\infty \exp[-2(\gamma + \ln \tau)]\mathrm{d}\tau : \quad \mathrm{Ei}(\tau) \to 0$$

$$= y(\infty) - e^{-2\gamma} \int_\tau^\infty \exp[-2\ln \tau]\mathrm{d}\tau \tag{1.20}$$

$$= y(\infty) - e^{-2\gamma} \int_\tau^\infty \tau^{-2}\mathrm{d}\tau$$

$$= y(\infty) - e^{-2\gamma}/\tau$$

where the Euler constant $\gamma = 0.5772$ and $y(\infty) = 0.747597$.

$y(\infty)$ is the numerical result from Eq. (1.19a) for $\tau \to \infty$.

Finally, we consider the size distribution $P(x)$ of the interval x where no particle exists after a long time ($\tau \to \infty$).

$$P(x) \propto - \mathrm{d}f(x, \infty)/\mathrm{d}x = S$$

and from Eq. (1.18a),

$$f(x, \tau) = \int_0^\tau \{-1 + 2e^{-\tau x}\}F(\tau)\mathrm{d}\tau \tag{1.18a}$$

$$-\mathrm{d}f(x, \tau)/dx = 2 \int_0^\tau \tau e^{-\tau x} \exp\left[-2 \int_0^\tau \{(1 - e^{-v})\mathrm{d}v/v\}\right]\mathrm{d}\tau$$

Defining S by

$$S = 2 \int_0^\infty \tau e^{-x\tau} \exp\left[-2 \int_0^\tau \{(1 - e^{-v})\mathrm{d}v/v\}\right]\mathrm{d}\tau$$

we obtain

$$\int_0^\infty S\,\mathrm{d}x = 2 \int_0^\infty \left[\int_0^\infty \tau e^{-x\tau}\mathrm{d}x\right] \exp\left[-2 \int_0^\tau \{(1 - e^{-v})\mathrm{d}v/v\}\right]\mathrm{d}\tau$$

$$= 2 \int_0^\infty \exp\left[-2 \int_0^\tau \{(1 - e^{-v})\mathrm{d}v/v\}\right]\mathrm{d}\tau = 2y(\infty)$$

Therefore,

$$P(x) = S/\left[\int_0^\infty S\,\mathrm{d}x\right]$$

$$= \{1/y(\infty)\} \int_0^\infty \tau e^{-x\tau} \exp\left[-2 \int_0^\tau \{(1 - e^{-v})/v\}\mathrm{d}v\right]\mathrm{d}\tau : 0 \leqq x \leqq 1 \tag{1.21}$$

The above equation is derived also by another method (Mackenzie, 1962). $P(x)$ diverges for $x \to 0$, because

$$\int_0^\tau \{(1 - e^{-v})/v\}\mathrm{d}v = \gamma + \ln \tau \tag{1.19b}$$

and substitution of Eq. (1.19b) into Eq. (1.21) gives

$$P(x) = \{1/y(\infty)\} \int_0^\infty \tau e^{-x\tau} \exp[-2(\gamma + \ln \tau)] d\tau$$

$$= \{1/y(\infty)\} \exp[-2\gamma] \int_0^\infty \tau e^{-x\tau} \exp[-2 \ln \tau] d\tau$$

$$= \{1/y(\infty)\} \exp[-2\gamma] \int_0^\infty \{\tau e^{-x\tau}/\tau^2\} d\tau$$

$$= \{1/y(\infty)\} \exp[-2\gamma] \int_0^\infty \{e^{-x\tau}/(x\tau)\} d(x\tau) \qquad (1.22)$$

$$= \{1/y(\infty)\} \exp[-2\gamma] \int_{\tau x}^\infty \{e^{-v}/v\} dv : \quad \tau x \fallingdotseq 0$$

$$= \{1/y(\infty)\} \exp[-2\gamma]\{\gamma + \ln \tau x + \cdots\}$$

$$= \{1/y(\infty)\} \exp[-2\gamma]\{\ln x + \gamma + \ln \tau + \cdots\}$$

This is because $x \to 0$ is equivalent to $\tau \to 0$ in the equation, so that the integral–exponential function diverges.

Equation (1.21) is shown in Figure 1.3 compared to the computer simulation experiments. The analytical solution is useful in considering the accuracy of the computer experiments.

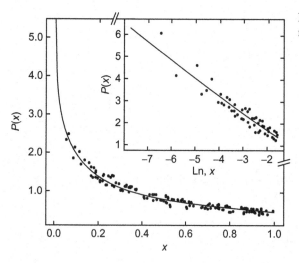

Figure 1.3 $P(x)$ versus x relation, Eq. (1.21).

References

Cohen, E.R., and Reiss, H., Kinetics of reactant isolation I. One-dimensional problems, J. Chem. Phys., vol. 38, 680 (1963).

Feder, J., Random sequential adsorption, J. Theor. Biol., vol. 87, 237 (1980).

Gonzalez, J.J., Hemmer, P.C., and Høye, J.S., Cooperative effects in random sequential polymer reactions, Chem. Phys., vol. 3, 228 (1974).

Hinrichsen, E.L., Feder, J., and Jøssang, T., Geometry of random sequential adsorption, J. Stat. Phys., vol. 44, 793 (1986).

Mackenzie, J.K., Sequential filling of a line by intervals placed at random and its application to linear adsorption, J. Chem. Phys., vol. 37, 723 (1962).

Rodgers, G.J., Random sequential adsorption of different-sized k-mers on a line, Phys. Rev. A, vol. 45, 3432 (1992).

2 Spatial Structure of Random Dispersion of Equal Spheres in Two Dimensions

2.1 Outline of Computer Simulation Experiments

The spatial structure of the random dispersion of equal spheres in two dimensions has been considered by several experimental methods, depending on the particle concentration. The methods are summarized in Table 2.1.

The random sequential adsorption (RSA) method (a) is the most fundamental and important, and has been widely used as a model of the adsorption phenomenon of gases onto the solid surface. The maximum possible value of the coverage fraction is $y = 0.5472 \pm 0.0002$, as reported by Hinrichsen et al. (1990). They improved the RSA method to make denser packing by introducing "the compressing operation," as explained in Table 2.1(b); the maximum possible coverage fraction becomes $y = 0.772 \pm 0.002$ that is close to the random loose packing density $y = 0.78$. The random closest packing density is $y = 0.82$ (Berryman, 1983), and Hinrichsen et al. (1990) reported the theoretical result of $y = \pi^2/12 = 0.822$.

Bennett (1972) proposed a model of simulating the growth of nuclei in aggregates and made the random packing of 1000 disks by the method in Table 2.1(c). The packing density is $y = 0.839$ in the central region, while $y = 0.823$ in the outer surface region. In other words, the central region consists of regular arrangement, while the outer tends to the random (Gotoh, 1993).

The Monte Carlo method by Metropolis et al. (1953) explained in Table 2.1(d) is widely used for simulating atomic and/or molecular systems in the statistical thermodynamics study, and the computer experiment is helpful for studying the physical/chemical properties of matter. Fraser (Fraser et al., 1990; Fraser, 1991) reported the computer experiment of the hard disk system.

Random packing can be made from the regular packing as explained in Table 2.1(e) and (f), though the disk arrangement is different from those of other methods.

Particulate Morphology. DOI: 10.1016/B978-0-12-396974-3.00002-3

Table 2.1 Computer Experiment Methods in Two Dimensions

(a) Random sequential adsorption (RSA) model

Equal disks are placed one by one randomly in a square region without overlap. A periodic boundary condition is adopted by assuming existence of the same square regions around the central one under consideration so as to remove the wall effect. If an inserting disk overlaps an already-placed one, it is disregarded and another disk is tried. In the case of no overlap, the disk is fixed in the position. The insertion of the disks is continued until the area fraction y of the square region covered with the disks reaches a prescribed value. The maximum possible coverage fraction is $y = 0.5472$.

(b) Random packing

The above RSA model is modified to make denser coverage fraction by "the compressing operation." In a random dispersion made by the RSA method, we consider the Voronoi polygons: each central disk center is connected with all neighboring disk centers, the center-to-center distances are bisected to give polygonal areas including the central disk, and the most internal polygon is the Voronoi polygon, that is, Voronoi cell. We consider the largest possible circle in each Voronoi cell and move the disk center to the circle center. After moving all disks, we expand all disks uniformly until any two neighboring disks are touching. The repetition of the compressing operation gives the maximum possible coverage fraction of $y = 0.772$.

(c) Random aggregate

A set of triangular contacting disks is placed at a center and equal disks fall one by one from random distant directions to the center. Each disk rolls on the existing disks down to the center until stably supported by two disks.

(d) Monte Carlo method

Consider a square region covered with equal disks at a prescribed area fraction. The disk arrangement is usually regular at the beginning of the simulation for simplicity. The periodic boundary condition explained in the RSA model (a) is adopted. The total energy U_i by the interaction between the disk i and all neighboring disks is calculated. And the disk i is moved to the point j by $k_x\delta$ in x-direction and $k_y\delta$ in y-direction, where k_x and k_y are the random number between -1 and $+1$ and δ is the maximum allowable step of the movement, that is, one-fifth of the disk diameter, for example. After the movement to the point j, the total energy U_j by the interaction between the disk j and all neighboring disks is calculated. If $U_j \leqq U_i$, the disk i is fixed at the point j and we forward to the next movement; if $U_j > U_i$, the movement is allowed with the probability $p = \exp[-(U_j - U_i)/(k_B T)]$. In other words, another random number r is introduced in the range $0 < r < 1$. And if $r \leqq p$, the disk i is fixed at the point j; if $r > p$, the disk i is returned to the original point and we forward to the movement of the next disk, where k_B is the Boltzmann constant and T is the absolute temperature.

The above movement is continued for all disks repeatedly until reaching the equilibrium state. In the case of the hard disks, $p = 0$ for overlap and the disk is returned to the original position; $p = 1$ for other cases and the movement is allowed.

(e) Loosest packing with random vacancies

The regular square packing of equal disks ($y = 0.785$) is considered and the disks are randomly removed one by one so as to make a prescribed area fraction covered with the disks.

(f) Closest packing with random vacancies

The regular hexagonal packing of equal disks ($y = 0.907$) is considered and the disks are randomly removed one by one so as to make a prescribed area fraction covered with the disks.

2.2 Structure of Random Dispersion

Aboav (1985a,b) reported a series of research on the geometric structure of random dispersion of points, in which he discussed several mosaic figures including the single point. The most well known is the Voronoi polygon, that is, the Voronoi cell. As depicted in Figure 2.1, a central particle or a central point in the case of size zero is connected with all neighboring ones and the lines are bisected to make polygons. The most inner one is the Voronoi polygon that contains only one particle or the single point. The Voronoi polygons can cover the whole region completely so that it is called the cell. The average shape of the cells becomes hexagonal and the shape distribution was studied in detail by Hinrichsen et al. (1986), (Fraser, 1991; Fraser et al., 1990), and Boots (1987).

Lado (1968) and Steele (1976) studied the radial distribution function for the random dispersion of equal disks. Feder (1980), Hinrichsen et al. (1986), and Hoover (1979) reported the size distribution of pores in the system.

This section outlines various research results on the size distribution of the Voronoi cell that is the minimum basic unit for considering the local structure of the particle dispersion. The regularity r of the cells is defined by

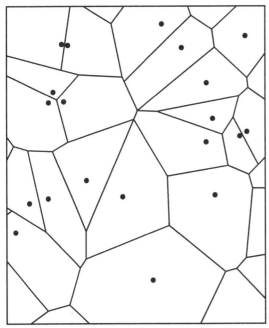

Figure 2.1 Illustration of Voronoi polygon.

$$r = (\langle a \rangle / s)^2; \quad s^2 = \langle (a - \langle a \rangle)^2 \rangle \tag{2.1}$$

where a is the area of single cell, $\langle a \rangle$ is the average of a, and s^2 is the variance. Normalizing by $x = a/\langle a \rangle$ and using the probability distribution function $P(x)$, the regularity becomes as follows:

$$r = 1 / \int_0^\infty (x - 1)^2 P(x) \mathrm{d}x \tag{2.2}$$

The area fraction covered by the disks is zero in the case of the point particles without area, and $P(x)$ obeys the gamma distribution as explained by Di Cenzo and Wertheim (1989) and Weaire et al. (1986):

$$P(x) = r^r x^{r-1} \, \mathrm{e}^{-rx} / \Gamma(r) \tag{2.3}$$

where $\Gamma(\ )$ is the gamma function. Equation (2.3) is called the Schultz distribution and is determined only by the regularity r as depicted in Figure 2.2 (Miller and Torquato, 1989).

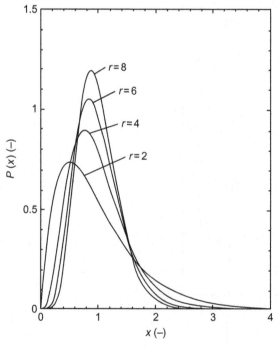

Figure 2.2 Gamma distribution function (Eq. (2.3)).

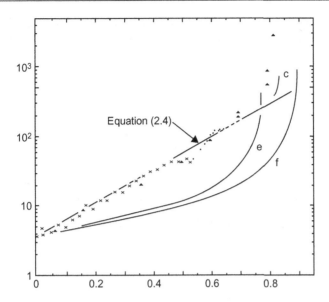

Figure 2.3 Regularity r versus coverage y relation $\times$: (a); $\bullet$ and $-$: (b); $\blacktriangle$: (d); (a)–(f) correspond to Table 2.1.

Figure 2.3 shows the regularity versus coverage relation obtained for various random dispersions in Table 2.1. The saturation point may exist at $y = 0.5472$ for the case (a) and at $y = 0.772$ for the case (b), where $dr/dy = 0$, and the coverage y does reach its upper limit for each simulation method.

The RSA model (a) and its compaction model (b) as a whole give the straight line as depicted in Figure 2.3, yielding (Gotoh, 1993):

$$r = 3.6 \exp[5.4y] \tag{2.4}$$

where $r = 3.6$ at $y = 0$ is close to the theoretical value $r = 2\pi/\sqrt{3}$ as derived by Weaire et al. (1986).

It may be said from Figure 2.3 that Eq. (2.4) corresponds to the most random dispersion; below which corresponds to the system with local voids; and above which corresponds to the system with local regular structures.

2.3   Measurement of Dispersed State of Disk Particles

Mechanical and physical properties of new composite materials and fine ceramics depend on the dispersed state of the discontinuous phase. A mixing operation exists and plays important roles in various chemical processes, where the local particle concentration is measured to evaluate the dispersed state of particles. It is usually impossible for one to know the absolute state of particle dispersion, because the concentration measurement depends on the probe size. One can only recognize its variation from place to place in the container. Microscopic methods are available for evaluation of the dispersed state of particles. The Voronoi polygonal analysis is the most detailed method, where the coordinates of particle positions are measured, and the Voronoi tessellation is conducted to obtain the distribution of cell areas. Although the method can provide precise data concerning particle dispersion, it has the disadvantage of being time consuming. Hence, a conventional simple method is proposed in this section to obtain the measure of regularity from the measurement of variation in the local particle concentrations by a probe of adjustable size. Evaluation of the state of particle dispersion therefore becomes faster for TV monitoring systems.

The system is supposed to consist of N equal squares, the size of which is equal to the particle diameter. M equal particles are distributed randomly into N cells. The particle concentration is measured by the probe of n cells. If m of n cells are filled with the particles, the measured local number concentration is $c_1 = m/n$ and its variance becomes as follows (cf. Section 5.2):

$$\sigma^2 = c(1-c)/n, \quad c = M/N, \quad n \ll N \tag{2.5}$$

And the regularity becomes as follows:

$$R = (c/\sigma)^2 = cn/(1-c) \tag{2.6}$$

where R depends on the probe size n, while the regularity r obtained from the Voronoi tessellation method is the most microscopic and independent of the probe size. However, the Voronoi tessellation method requires the coordinates of particle positions and it is time consuming. Accordingly, we consider the concentration measurement by choosing the probe size so as to become $R = r$, yielding

$$cn/(1-c) = r$$

where r is given by Eq. (2.4) in the Section 2.2, $c = 4y/\pi$, and y is the bulk-mean particle area fraction, that is, the coverage. Hence,

$$n = 3.6(4y/\pi - 1)\exp[5.4y] \tag{2.7}$$

In summary, a simple conventional method has been discussed in this section to estimate the regularity of particle dispersion from the measurement of variation in local particle concentrations, and the following procedure is proposed.

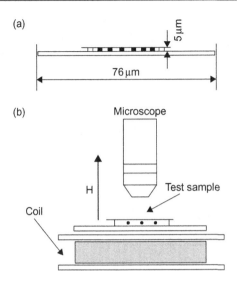

Figure 2.4 Main section of experimental apparatus: (a) test sample and (b) test section.

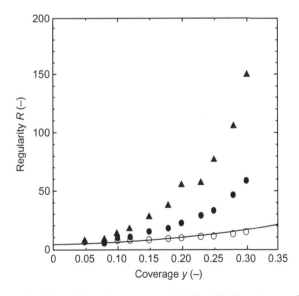

Figure 2.5 Regularity R in relation to coverage y. (○) For initial random configuration; (●) for magnetic field $H = 25,200$ (A/m); and (▲) for $H = 40,000$ (A/m). Solid curve is $R = 3.6 \exp[5.4y]$.

(a)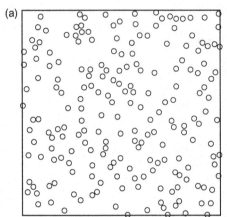

Figure 2.6 Particle arrangement. $y = 0.1$ $(-)$. (a) Initial random configuration; (b) final configuration, magnetic field $H = 25,200$ (A/m); and (c) final configuration, $H = 40,000$ (A/m).

(b)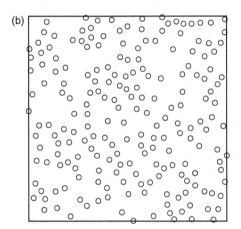

(c)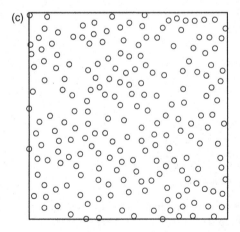

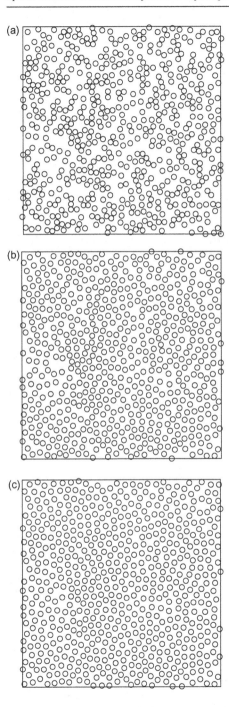

Figure 2.7 Particle arrangement. $y = 0.3$ ($-$).
(a) Initial random configuration; (b) final
configuration, magnetic field $H = 25{,}200$ (A/m);
and (c) final configuration, $H = 40{,}000$ (A/m).

1. First, the spatial variation of local particle concentration is measured by a probe large enough to obtain the coverage y, that is, the bulk-mean fractional area of particles.
2. The probe size n is adjusted to the value calculated from Eq. (2.7).
3. Variation in the local particle concentration is measured again by the probe. Then, we obtain the regularity that is equal to the value obtained from the Voronoi tessellation method.

In this way, the evaluation of the dispersed state of particles becomes faster for TV monitoring systems (Itoh et al., 1995).

Application

Particles confined in a thin horizontal layer of magnetic fluid are uniformly dispersed or form chain-like clusters depending on the direction of the applied magnetic field. Similar phenomena occur in electrorheological fluids. This can be a model for studying the phase change from solid to liquid, as well as a visible model of colloidal systems.

The conventional method proposed above is applied to evaluate the particle dispersion, where silica particles of diameter $d = 3.0$ µm are confined in a thin layer (5 µm thick) of magnetic fluid, and the magnetic field is applied in the vertical direction. Resulting particle movements are investigated to examine effects of the magnetic field and the bulk-mean particle concentration on the regularity of the particle dispersion. As the magnetic field and the particle concentration increase, the particle arrangement is found to become more regular due to the repulsive force among particles as shown in Figures 2.4–2.7 (Horizoe et al., 1995).

References

Aboav, D.A., The arrangement of cells in a net, III, Metallography, vol. 17, 383 (1985a).

Aboav, D.A., The arrangement of cells in a net, IV, Metallography, vol. 18, 129 (1985b).

Bennett, C.H., Serially deposited amorphous aggregates of hard spheres, J. Appl. Phys., vol. 43, 2727 (1972).

Berryman, J.G., Definition of dense random packing. Advances in the Mechanics and the Flow of Granular Materials, vol. 1, pp. 1–18 (ed. by M. Shahinpoor), Trans. Tech, Germany (1983).

Boots, B.N., Edge length properties of random Voronoi polygons, Metallography, vol. 20, 231 (1987).

Di Cenzo, S.B., and Wertheim, G.K., Monte Carlo calculation of the size distribution of supported clusters, Phys. Rev. B, vol. 39, 6792 (1989).

Feder, J., Random sequential adsorption, J. Theor. Biol., vol. 87, 237 (1980).

Fraser, D.P., Voronoi statistics of hard-core systems, Mater. Charact., vol. 26, 73 (1991).

Fraser, D.P., Zuckermann, M.J., and Mouritsen, O.G., Simulation technique for hard-disk models in two dimensions, Phys. Rev. A, vol. 42, 3186 (1990).

Gotoh, K., Comparison of random configurations of equal disks, Phys. Rev. E, vol. 47, 316 (1993).

Hinrichsen, E.L., Feder, J., and Jøssang, T., Geometry of random sequential adsorption, J. Stat. Phys., vol. 44, 793 (1986).

Hinrichsen, E.L., Feder, J., and Jøssang, T., Random packing of disks in two dimensions, Phys. Rev. A, vol. 41, 4199 (1990).

Hoover, W.G., Exact hard-disk free volumes, J. Chem. Phys., vol. 70, 1837 (1979).

Horizoe, M., Itoh, R., and Gotoh, K., Uniform dispersion of fine particles in a magnetic fluid and its evaluation, Adv. Powder Technol., vol. 6, no. 2, 139 (1995).

Itoh, R., Horizoe, M., and Gotoh, K., A method for measuring two-dimensional dispersed state of particles, Adv. Powder Technol., vol. 6, no. 2, 81 (1995).

Lado, F., Equation of state of the hard-disk fluid from approximate integral equations, J. Chem. Phys., vol. 49, 3092 (1968).

Metropolis, N., Rosenbluth, M.N., Teller, A.H., and Teller, E., Equation of state calculations by fast computing machines, J. Chem. Phys., vol. 21, 1087 (1953).

Miller, C.A., Diffusion-controlled reactions among spherical traps: Effect of polydispersity in trap size, Phys. Rev. B, vol. 40, 7101 (1989).

Steele, W.A., Theory of monolayer physical adsorption I. Flat surface, J. Chem. Phys., vol. 65, 5256 (1976).

Weaire, D., Kermode, J.P., and Wejchert, J., On the distribution of cell areas in a Voronoi network, Philos. Mag. B, vol. 53, L101 (1986).

3 Preliminary Mathematics

Mathematical procedures for dealing with the radial distribution function in the next chapter are explained below.

3.1 Laplace Transform and Inversion Formula

3.1.1 Laplace Transform

$f(t)$ is defined for the region $0 < t < \infty$ and if the following integral

$$f(p) = \int_0^\infty f(t)e^{-pt}\,dt$$

exists for the complex number p, it is called the Laplace integral and expressed by $f(p) = L[f(t)]$.

The inversion of the Laplace transform is expressed by

$$f(t) = L^{-1}[f(p)] = (2\pi i)^{-1} \int_{b-i\infty}^{b+i\infty} f(p)e^{pt}\,dp$$

Proof

$$I = (2\pi i)^{-1} \int_{b-iT}^{b+iT} f(p)e^{pt}\,dp : \quad p = b + iu, \quad dp = i\,du$$

$$= (2\pi)^{-1} \int_{-T}^{+T} f(b+iu)e^{(b+iu)t}\,du : \quad f(p) = \int_0^\infty f(x)e^{-px}\,dx$$

$$= (2\pi)^{-1} \int_{-T}^{+T} e^{(b+iu)t} \left[\int_0^\infty f(x)e^{-(b+iu)x}\,dx \right] du$$

$$= (2\pi)^{-1} \int_0^\infty f(x)\,dx \int_{-T}^{+T} e^{(b+iu)(t-x)}\,du$$

$$= \int_0^\infty f(x)\,dx \ \pi^{-1} e^{b(t-x)} \int_0^T \cos(t-x)u\,du$$

Particulate Morphology. DOI: 10.1016/B978-0-12-396974-3.00003-5

(cont'd)

$$= \int_0^\infty f(x)\delta(t-x)\mathrm{d}x : \quad T \to \infty$$

$$= [f(t+0) + f(t-0)]/2 : \quad \text{for discontinuous case}$$

$$= f(t) : \quad \text{for continuous case}$$

(Reference): Delta function $\delta(t-x)$

$$L[\delta(t-x)] = \int_0^\infty \delta(t-x)\mathrm{e}^{-pt}\,\mathrm{d}t = \mathrm{e}^{-px}$$

$$\delta(t-x) \quad = (2\pi\mathrm{i})^{-1} \int_{b-\mathrm{i}\infty}^{b+\mathrm{i}\infty} \mathrm{e}^{-px}\,\mathrm{e}^{pt}\,\mathrm{d}p$$

$$= \pi^{-1}\,\mathrm{e}^{b(t-x)} \int_0^T \cos(t-x)u\,\mathrm{d}u : \quad T \to \infty$$

3.1.2 Inversion of Laplace transform

In the following relation,

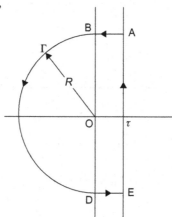

$$f(t) = L^{-1}[f(p)] = (2\pi\mathrm{i})^{-1} \int_{b-\mathrm{i}\infty}^{b+\mathrm{i}\infty} f(p)\mathrm{e}^{pt}\,\mathrm{d}p$$

The poles of the complex function $f(p)$ are assumed to be $p_1, p_2, p_3, \ldots, p_n$ and their real parts are smaller than a large real number b: $\mathrm{Re}(p_n) < b$ for all p_n.

Consider a closed path C composed by the lines DE, EA, and AB, and the semi-circle Γ as depicted in the figure. e^{pt} is analytic and nonzero, and hence the product, $e^{pt}f(p)$, has the same poles as those of $f(p)$.

Note the residue of $e^{pt}f(p)$ at $p = p_n$ by $\Phi_n(t)$ for a fixed t. According to the Cauchy theorem, the integral of $e^{pt}f(p)$ along the closed path around the poles p_1, $p_2, p_3, \ldots, p_n$ becomes $2\pi i \sum_{k=1}^{n} \Phi_k(t)$. Therefore, the inversion yields

$$f(t) = L^{-1}[f(p)] = \sum_{k=1}^{n} \mathrm{Res}(p_k)$$

3.1.3 Residue

When a is the isolated singular point of the complex function $f(z)$, i.e., $f(a) = 0$ in the complex number plane, consider the integral along a small circle C including the pole a at the center. In the case of the sufficiently small circle, it contains the single pole a only. Then the residue of $f(z)$ at $z = a$ is defined by

$$\mathrm{Res}(a) = (2\pi i)^{-1} \int_C f(z)\mathrm{d}z$$

For the isolated singular point a, the complex number function $f(z)$ can be expressed by

$$f(z) = \Phi(z)/(z - a) = A_{-1}(z - a)^{-1} + A_0 + A_1(z - a) + A_2(z - a)^2 + \cdots$$
$$\Phi(z) = A_{-1} + A_0(z - a) + A_1(z - a)^2 + A_2(z - a)^3 + \cdots$$

Hence,

$$\mathrm{Res}(a) = (2\pi i)^{-1} \int_C \Phi(z)/(z - a)\mathrm{d}z = \Phi(a) = A_{-1}$$

When $f(z) = 1/g(z)$,

$$g(z) = g'(a)(z - a) + (1/2)g''(a)(z - a)^2 + \cdots : \quad g'(a) \neq 0$$

and therefore

$$\mathrm{Res}(a) \quad = (2\pi i)^{-1} \int_C \mathrm{d}z/[(z - a)\{g'(a) + (1/2)g''(a)(z - a) + \cdots\}]$$
$$= 1/g'(a)$$

For the singular point a of the order n:

$$f(z) = A_{-n}(z - a)^{-n} + A_{-n+1}(z - a)^{-n+1} + A_{-n+2}(z - a)^{-n+2} + \cdots = \Phi(z)/(z - a)^n$$

$$\Phi(z) = A_{-n} + A_{-n+1}(z - a) + \cdots$$

Hence,

$$\mathrm{Res}(a) = (2\pi i)^{-1} \int_C f(z)\mathrm{d}z = (2\pi i)^{-1} \int_C \Phi(z)/(z-a)^n \, \mathrm{d}z$$
$$= \Phi^{(n-1)}(a)/(n-1)!$$

The complex number function $f(z)$ has the isolated singular points $a_1, a_2, \ldots, a_n$ in the region D of the closed path C, and their residues are expressed by $\mathrm{Res}(a_1)$, $\mathrm{Res}(a_2), \ldots, \mathrm{Res}(a_n)$. If $f(z)$ is regular in D and on C except of the above singular points, the following relation holds:

$$(2\pi i)^{-1} \int_C f(z)\mathrm{d}z = \mathrm{Res}(a_1) + \mathrm{Res}(a_2) + \cdots + \mathrm{Res}(a_n)$$

Proof

Small circles $C_1, C_2, \ldots, C_n$ are considered for each points $a_1, a_2, \ldots, a_n$. Then

$$\int_C f(z)\mathrm{d}z + \int_{-C_1} f(z)\mathrm{d}z + \int_{-C_2} f(z)\mathrm{d}z + \cdots + \int_{-C_n} f(z)\mathrm{d}z = 0$$

where the integral along the closed path is plus for the clockwise direction and minus for the counterclockwise. And therefore,

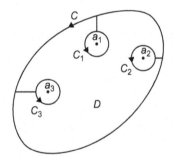

$$(2\pi i)^{-1} \int_C f(z)\mathrm{d}z = (2\pi i)^{-1} \int_{C_1} f(z)\mathrm{d}z$$
$$+ (2\pi i)^{-1} \int_{C_2} f(z)\mathrm{d}z + \cdots + (2\pi i)^{-1} \int_{C_n} f(z)\mathrm{d}z$$
$$= \mathrm{Res}(a_1) + \mathrm{Res}(a_2) + \cdots + \mathrm{Res}(a_n)$$

3.2   Fourier Transform and Spectral Density

3.2.1   Fourier Transform and Inversion Formula

The Fourier integral of $f(t)$ is defined for $-\infty < t < \infty$ as follows:

$$f(u) = \int_{-\infty}^{\infty} f(t)e^{-iut} \, dt$$

where u is a real number. The Fourier transform and its inversion are denoted, respectively, by $f(u) = F[f(t)]$ and $f(t) = F^{-1}[f(u)]$. Then

$$f(t) = F^{-1}[f(u)] = (2\pi)^{-1} \int_{-\infty}^{\infty} f(u)e^{iut} \, du$$

$$= (2\pi)^{-1} \int_{-\infty}^{\infty} \int_{-\infty}^{\infty} f(x)e^{-iux} \, dx \, e^{iut} \, du$$

$$= (2\pi)^{-1} \int_{-\infty}^{\infty} f(x)dx \int_{-\infty}^{\infty} e^{iu(t-x)} du$$

$$= \int_{-\infty}^{\infty} f(x)dx (2\pi)^{-1} \int_{-\infty}^{\infty} e^{iu(t-x)} du$$

$$= \int_{-\infty}^{\infty} f(x)dx \, \delta(t - x)$$

$$= f(t)$$

3.2.2   Spectral Density and Correlation Function

Consider the random variable $x(t)$ at time t and its truncation as follows:

$$\begin{aligned} x_T(t) \quad &= x(t) \quad : \quad |t| \leq T \\ &= 0 \quad : \quad |t| > T \end{aligned}$$

The Fourier transform of $x_T(t)$ is denoted by $A_T(f)$, and hence,

$$A_T(f) = \int_{-\infty}^{\infty} x_T(t)e^{-iut} \, dt \quad = \int_{-T}^{T} x_T(t)e^{-iut} \, dt$$

$$= \int_{-T}^{T} x_T(t)e^{-i2\pi ft} \, dt$$

The spectral density is defined by

$$S(f) = \lim_{T \to \infty} |A_T(f)|^2 / T$$

And the autocorrelation function is defined by

$$R_{xx}(\tau) = \lim_{T \to \infty} R_{xx}^T(\tau) = \lim_{T \to \infty} \int_{-\infty}^{\infty} x_T(t)x_T(t + \tau)dt/(2T)$$

The Fourier transform becomes as follows:

$$\int_{-\infty}^{\infty} R_{xx}^T(\tau)e^{-i2\pi f\tau}\, d\tau = \int_{-\infty}^{\infty} x_T(t)x_T(t + \tau)dt\, e^{-i2\pi f\tau}\, d\tau/(2T)$$

$$= \int_{-\infty}^{\infty} x_T(t)e^{i2\pi ft}\, dt \int_{-\infty}^{\infty} x_T(t + \tau)dt\, e^{-i2\pi f(t+\tau)}d\tau/(2T)$$

$$= A_T(-f)A_T(f)/(2T) = |A_T(f)|^2/(2T)$$

Hence, the spectral density of the autocorrelation function becomes

$$S_{xx}(f) = \lim_{T \to \infty} 2 \int_{-\infty}^{\infty} R_{xx}^T(\tau)e^{-i2\pi f\tau}\, d\tau$$

$$= 2 \int_{-\infty}^{\infty} R_{xx}(\tau)e^{-i2\pi f\tau}\, d\tau$$

$$= 4 \int_{0}^{\infty} R_{xx}(\tau)\cos(2\pi f\tau)d\tau$$

$$R_{xx}(\tau) = \int_{-\infty}^{\infty} S_{xx}(f)e^{i2\pi f\tau}df/2 = \int_{0}^{\infty} S_{xx}(f)\cos(2\pi f\tau)df$$

In the case of the cross-correlation function, the second variable function $y(t)$ is introduced.

$$y_T(t) = y(t) \quad : \quad |t| \leq T$$
$$= 0 \quad : \quad |t| > T$$

$$R_{xy}^T(\tau) = \int_{-\infty}^{\infty} x_T(t)y_T(t + \tau)dt/(2T)$$

$$\int_{-\infty}^{\infty} R_{xy}^T(\tau)e^{-i2\pi f\tau}\,d\tau = \int_{-\infty}^{\infty} x_T(t)e^{i2\pi ft}\,dt \int_{-\infty}^{\infty} y_T(t+\tau)e^{-i2\pi f(t+\tau)}\,d\tau/(2T)$$

$$= A_T(-f)B_T(f)/(2T)$$

Therefore,

$$S_{xy}(f) = 2\int_{-\infty}^{\infty} R_{xy}(\tau)e^{-i2\pi f\tau}\,d\tau$$

$$R_{xy}(\tau) = \int_{-\infty}^{\infty} S_{xy}(f)e^{i2\pi f\tau}\,df/2$$

Example

$$x(t) = A + B\sin(2\pi f^* t)$$

$$R_{xx}(\tau) = A^2 + [B^2\cos(2\pi f^*\tau)]/2$$

$$S_{xy}(f) = 4\int_0^{\infty} \{A^2 + [B^2\cos(2\pi f^*\tau)]/2\}\cos(2\pi f\tau)\,d\tau$$

$$= 4A^2\int_0^{\infty}\cos(2\pi f\tau)\,d\tau + 2B^2\int_0^{\infty}\cos(2\pi f^*\tau)\cos(2\pi f\tau)\,d\tau$$

$$= 2A^2\delta(f) + (B^2/2)[\delta(f-f^*) + \delta(f+f^*)]$$

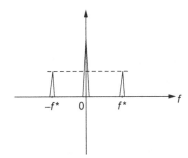

4 Radial Distribution Function

4.1 Definition of Radial Distribution Function

Consider a random assembly of equal spheres and the particle number density, that is, the number concentration $\rho(r)$ at the position r. For small volumes, dv at r and dv' at r', the number of particle pairs is expressed by $\rho^2(r,r')dv\,dv'$. The number of pairs made by one single particle in dv and all particles in dv' becomes as follows:

$$'\rho^2(r,r') = \rho^2(r,r')dv'/\rho(r)$$

The correlation coefficient is defined by

$$g(r,r') = \rho^2(r,r')/\{\rho(r)\rho(r')\}$$

In the case of uniform system $\rho(r) = \rho = N/V = $ constant so that $g(r,r')$ becomes the function of the distance $|r-r'| = R$ only. Hence,

$$g(R) = \rho^2(r,r')/\rho^2$$

The particle number concentration at the radial distance R from an arbitral particle center becomes

$$'\rho^2(r,r')/dv' = \rho^2(r,r')/\rho = \rho g(R) \tag{4.1}$$

The above relation defines the radial distribution function $g(R)$ as the magnification factor for the bulk-mean value ρ.

Particulate Morphology. DOI: 10.1016/B978-0-12-396974-3.00004-7

4.2 Ornstein–Zernike Equation

Equation (4.1) is rewritten as follows:

$$
'\rho^2(r,r')/dv' = \rho g(R)
$$
$$
= \rho + \rho h(R) = \rho\{1 + h(R)\}
\tag{4.2}
$$

where $h(R)$ is the fractional deviation of the local number density from the bulk-mean value.

The direct correlation function $c(R)$ is introduced such that the local number density deviates from ρ to $\rho c(R)$ by the direct effect of a central particle placed at $R = 0$. Here, consider a small volume $\{dR'\}$ at another radial position R' and the indirect effect of the central particle via the particles in $\{dR'\}$ onto the position R is denoted by $\rho I(R)$ as follows:

$$
\rho h(R) = \rho c(R) + \rho I(R)
\tag{4.3}
$$

There are $\rho\{dR'\}$ particles in the small volume at the radial position R', one of which affects the local number density at R by the amount of deviation $\rho h(R - R')$ and hence,

$$
\rho I(R) = \iiint \rho c(R')\{dR'\} \times \rho h(R - R')
\tag{4.4}
$$

Accordingly, the fractional deviation of the particle number concentration becomes

$$
h(R) = c(R) + \rho \iiint c(R')h(R - R')\{dR'\}
\tag{4.5}
$$

where $\{dR'\}$ is the small volume at the position R' and

$$
h(R) = g(R) - 1; \quad h(R) = 0 \quad \text{for } R \to \infty
$$

Eq. (4.5) is called the Ornstein–Zernike equation and defines the direct correlation function $c(R)$.

4.3 Solving Procedure

Assuming the isotropic system, we choose a flat plate of the unit thickness consisted of the origin 0 and two positions R and R'. The deviation of the particle number in the small volume $\{dR'\}$ at R' from the bulk-mean value is

$$\rho R'\, d\theta\, dR' c(R')$$

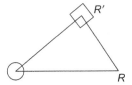

Choosing a single particle in the small volume as the center, the local particle number density is considered at the distance $(R - R')$, yielding $\rho h(R - R')$. And the substitution of $dZ = 1$ into Eq. (4.4) becomes

$$\rho I(R) = \int\int \rho R'\, d\theta\, dR' c(R') \times \rho h(R - R')$$

The small volume at R is $dV = R\, d\theta\, dR$, the small volume at the distance $(R - R')$ from the center point R' is $dV' = (R - R')d\psi dR$, and $\{dR'\} = R' d\theta dR'$. Therefore, from Eq. (4.5) we obtain

$$R\, dRh(R) = R\, dRc(R) + \rho \int\int R'\, dR' c(R') \times (R - R')d\psi\, dRh(R - R') \qquad (4.6)$$

By use of the following relations

$$dH(R) = 2\pi Rh(R)dR$$

$$dQ(R) = 2\pi Rc(R)dR \qquad (4.7)$$

Eq. (4.5) in the three-dimensional expression is simplified to become one-dimensional one for the case of the isotropic system as follows:

$$dH(R) = dQ(R) + \rho \int\int dQ(R')dH(R - R')d\psi/(2\pi)$$

$$= dQ(R) + \rho \int dQ(R')dH(R - R') \qquad (4.8)$$

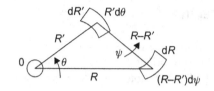

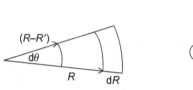 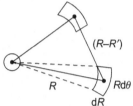

4.4 Analytic Solution by Percus–Yevick Approximation

In the previous section, Eq. (4.8) is obtained for the one-dimensional expression of the Ornstein–Zernike equation as follows:

$$dH(R) = dQ(R) + \rho \int dQ(R')dH(R-R')$$

By the definition of Eq. (4.7),

$$2\pi Rh(R) = dQ(R)/dR + \rho \int dQ(R')2\pi(R-R')h(R-R')$$

where $h(R) = g(R) - 1$.

In the region $R < D$ (=the particle diameter), $g(R) = 0$ and $h(R) = -1$. The Percus–Yevick approximation assumes $c(R) = 0$ for $R \geqq D$. Then the integration with respect to R' for the region $0-\infty$ yields

$$2\pi Rh(R) = dQ(R)/dR + 2\pi\rho \int_0^D dQ(R') \, (R - R')h(R-R') \qquad (4.9)$$

$Q(R)$ under consideration is only for the region $R = 0$ to D where $h(R) = -1$.
Hence, from Eq. (4.9) we obtain

$$dQ(R)/dR = AR + B \qquad (4.10)$$

where

$$A = -2\pi \left\{ 1 + \rho \int_0^D dQ(R') \right\}$$

$$B = 2\pi\rho \int_0^D R' \, dQ(R') \tag{4.11}$$

And therefore,

$$Q(R) = AR^2/2 + BR + K \tag{4.12}$$

The constant K can be determined from the conditions $Q(D) = 0$, Eqs. (4.11) and (4.12) as follows:

$$A = -2\pi(1 + 2\phi)/(1 - \phi)^2$$

$$B = 3\pi D\phi/(1 - \phi)^2$$

$$K = \pi D^2/(1 - \phi) \tag{4.13a}$$

where $\phi = \pi\rho D^3/6$: bulk-mean particle volume fraction.
From Eqs. (4.9) and (4.10),

$$G(R) - R - (AR + B)/(2\pi) + \rho \int_0^D Q(R')\{G(R - R') - (R - R')\}dR' = 0$$

where

$$G(R) = Rg(R)$$

The Laplace transform of the above relation becomes

$$G(p) - 1/p^2 - \int_0^D (AR + B)\exp[-pR]dR/(2\pi) +$$

$$\rho \int_0^D Q(R')\{G(p)\exp[-R'p] - 1/p^2 + R'/p\}dR' = 0$$

where p is the parameter. The integration with respect of R' gives

$$G(t) = tL(t)/\{12\phi L(t) + S(t)\exp(t)\}$$
$$= t[L(t)/\{S(t)\exp(t)\}]/[1 + 12\phi L(t)/\{S(t)\exp(t)\}]$$

Using $x/(1 + x) = \sum_{n=1}^{\infty} (-1)^{n-1}x^n$

$$G(t) = \sum_{n=1}^{\infty} (-12\phi)^{n-1}t\{L(t)/S(t)\}^n\exp(-tn)$$

where

$$t = Dp$$

$$L(t) = (1 + \phi/2)t + 1 + 2\phi$$

$$S(t) = (1 - \phi)^2 t^3 + 6\phi(1 - \phi)t^2 + 18\phi^2 t - 12\phi(1 + 2\phi) \qquad (4.13b)$$

$$\phi = \pi\rho D^3/6 : \text{bulk} - \text{mean particle volume fraction}$$

Introducing the unit function $H(x)$:

$$g(x) = \sum_n H(x - n)g_n(x)$$

and the Laplace transform:

$$g(t) = \sum_n g_n(t)\exp(-nt) \qquad (4.14)$$

$$G(t) = \sum_n tg_n(t)$$

$$tg_n(t) = (-12\phi)^{n-1}t\{L(t)/S(t)\}^n\exp(-tn)$$

Therefore,

$$xg_n(x) = [(-12\phi)^{n-1}/(n - 1)!]$$
$$\times \sum_{ti} \lim_{t \to ti} d^{n-1}/dt^{n-1}[t(t - t_i)^n\{L(t)/S(t)\}^n]\exp\{t(x - n)\} \qquad (4.15a)$$

where Σ is the sum for the three roots t_i ($i = 0,1,2$) of $S(t) = 0$.

$$t_i = \{-2\phi + (2\phi f)^{1/3}(y_+ j^i + y_- j^{-i})\}/(1 - \phi) \qquad (4.15b)$$

and

$$f = 3 + 3\phi - \phi^2$$

$$y_+ = [1 + (1 + 2\{\phi^2/f\}^2)^{1/2}]^{1/3}$$

$$y_- = [1 - (1 + 2\{\phi^2/f\}^2)^{1/2}]^{1/3}$$

$$j = \{-1 + (-3)^{1/2}\}/2$$

Throop and Bearman (1965) and Mandel et al. (1970) reported the numerical chart of the radial distribution function $g(x)$ by the Percus−Yevick approximation for the region of the dimensionless distance $x = R/D = 0-6$.

A. $xg_n(x)$, Eq. (4.15), by Smith and Henderson (1970):

$$
\begin{aligned}
g(x) &= 0 &&: 0 < x < 1 \\
&= g_1(x) &&: 1 \leqq x \leqq 2 \\
&= g_1(x) + g_2(x) &&: 2 \leqq x \leqq 3 \\
&= g_1(x) + g_2(x) + g_3(x) &&: 3 \leqq x \leqq 4 \\
&\vdots
\end{aligned}
$$

$$xg_1(x) = \sum_{i=0}^{2} a_0^i \exp\{t_i(x - 1)\}$$

$$xg_2(x) = \sum_{i=0}^{2} b_0^i \{b_1^i + b_2^i(x - 2)\}\exp\{t_i(x - 2)\}$$

$$xg_3(x) = \sum_{i=0}^{2} c_0^i \{c_1^i + c_2^i(x - 3) + c_3^i(x - 3)^2\}\exp\{t_i(x - 3)\}$$

$$xg_4(x) = \sum_{i=0}^{2} d_0^i \{d_1^i + d_2^i(x - 4) + d_3^i(x - 4)^2 + d_4^i(x - 4)^3\}\exp\{t_i(x - 4)\}$$

where the superscript i corresponds to Eq. (4.15b). For example, a_0^i is obtained from the following a_0 by substituting $t = t_i$.

$$L = L(t), \quad L_1 = dL(t)/dt$$

$$S = S(t), \quad S_1 = dS(t)/dt, \quad S_2 = d^2S(t)/dt^2, \quad S_3 = d^3S(t)/dt^3$$

$$A_1 = (L/S_1)^2, \quad A_2 = 15S_2^2 - 4S_1S_3, \quad A_3 = L + 4tL_1, \quad A_4 = LS_2/S_3, \quad A_5 = 2L + 3tL_1$$

and

$$a_0 = tL/S_1$$

$$b_0 = -12\phi L/S_1^2, \quad b_1 = (1 - tS_2/S_1)L + 2tL_1, \quad b_2 = tL$$

$$c_0 = 72\phi^2 L/S_1^3, \quad c_1 = A_1t(3S_2^2 - S_1S_3) - 3L(S_2/S_1)(L + 3tL_1) + 6L_1(L + tL_1)$$

$$c_2 = \{6tL_1 + L(2 - 3tS_2/S_1)\}L, \quad c_3 = tL^2$$

$$d_0 = -288\phi^3 L/S_1^4$$

$$d_1 = 5t(L/S_1)^3 S_2(2S_1S_3 - 3S_2^2) + A_1A_2A_3 - 24L_1A_4A_5 + 12L_1^2(3L + 2tL_1)$$

$$d_2 = \{tA_1A_2 - 12(A_3A_4 - L_1A_5)\}L, \quad d_3 = (3A_3 - 6tA_4)L^2, \quad d_4 = tL^3$$

B. $g(R)$ at the state of contact: $R = D$

The particle assembly can play an important role as the fundamental model gas in the statistical thermodynamics, where the equation of state is derived for the hard-sphere gas, as explained in later chapters by use of $g(R)$ at $R = D$.

From Eq. (4.13b),

$$S(t) = (1 - \phi)^2 t^3 + 6\phi(1 - \phi)t^2 + 18\phi^2 t - 12\phi(1 + 2\phi) = 0$$

and $t_i = t_0, t_1, t_2$ are obtained in Eq. (4.15b).

$$a_0^i = [(1 + \phi/2)t_i^2 + (1 + 2\phi)t_i]/[3(1 - \phi)^2 t_i^2 + 12\phi(1 - \phi)t_i + 18\phi^2]$$

The last term $18\phi^2$ is expressed by t_0, t_1, and t_2, yielding

$$(1 - \phi)^2 a_0^i = [(1 + \phi/2)t_i^2 + (1 + 2\phi)t_i]/[3t_i^2 - 2(t_0 + t_1 + t_2)t_i + t_0 t_1 + t_1 t_2 + t_2 t_0]$$

In the summation for $i = 0,1,2$, the denominator is unified to become $(t_0 - t_1)$ $(t_1 - t_2)(t_2 - t_0)$ and hence

$$(1 - \phi)^2 [a_0^0 + a_0^1 + a_0^2](t_0 - t_1)(t_1 - t_2)(t_2 - t_0) = (1 + \phi/2)(t_0 - t_1)(t_1 - t_2)(t_2 - t_0)$$

Accordingly, the radial distribution function at $R = D$ becomes

$$g(x = R/D) = g_1(1) = a_0^0 + a_0^1 + a_0^2 = (1 + \phi/2)/(1 - \phi)^2$$

4.5 Radial Distribution Function of Multisized Particle System

The Ornstein–Zernike equation is generalized for applying to the multisized particle system. Consider the spherical particle group a, b, i,... of the diameter D_a, D_b, D_i,... and the particle number density ρ_a, ρ_b, ρ_i,... respectively. The Ornstein–Zernike equation (4.5) is generalized to become

$$h_{ab}(r) = c_{ab}(r) + \sum_i \rho_i \int\int\int c_{ai}(r')h_{ib}(r - r')dr' \tag{4.16}$$

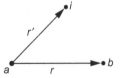

Σ_i denotes the summation for the subscript i, and the particle positions are depicted in the above figure, where the center of particle a is chosen as the origin. Using the following relations,

$$h_{ab}(\Omega) = (\rho_a \rho_b)^{1/2} \int\int\int e^{-j(\Omega \cdot r)} h_{ab}(r)dr$$

$$c_{ab}(\Omega) = (\rho_a \rho_b)^{1/2} \int\int\int e^{-j(\Omega \cdot r)} c_{ab}(r)dr \tag{4.17}$$

the Fourier transform of Eq. (4.16) becomes

$$h_{ab}(\Omega) = c_{ab}(\Omega) + \sum_i c_{ai}(\Omega)h_{ib}(\Omega) \tag{4.18}$$

where Ω is the three-dimensional parameter.

Because various pairings are possible among particles a, b, i,..., Eq. (4.18) can be expressed in terms of the matrix, and hence,

$$(\underline{1} - \underline{c}(\Omega))(\underline{1} + \underline{h}(\Omega)) = \underline{1} \tag{4.19}$$

where $\underline{1}$ is the unit matrix.

Following Baxter (1970), the matrix $\underline{Q}(\Omega)$ and its conjugate-transposed matrix $\underline{Q}^*(\Omega)$ are introduced by Eqs. (4.20a) or (4.20b):

$$(\underline{1} - \underline{c}(\Omega)) = \{\underline{1} + \underline{Q}^*(-\Omega)\}\{\underline{1} + \underline{Q}(\Omega)\} \tag{4.20a}$$

or

$$\underline{c}(\Omega) + \underline{Q}(\Omega) + \underline{Q}^*(-\Omega) + \underline{Q}^*(-\Omega)\underline{Q}(\Omega) = \underline{0} \tag{4.20b}$$

In the isotropic case under consideration, all terms become dependent only of the absolute value $|\Omega| = \omega$ in place of Ω, and therefore Eq. (4.20b) is simplified to become a one-dimensional expression as follows:

$$C_{ab}(r) + Q_{ab}(r) + Q_{ba}(-r) + \sum_i n_i \int_{-\infty}^{\infty} Q_{ia}(-r')Q_{ib}(r - r')dr' = 0 \tag{4.21}$$

where

$$C_{ab}(r) = (\rho_a\rho_b)^{-1/2} \int_{-\infty}^{\infty} e^{i\omega r} c_{ab}(\omega)d\omega \tag{4.22}$$

In other words, the introduction of the unknown function $Q_{ab}(r)$ into Eq. (4.19) yields

$$\underline{h}(\Omega) + \underline{Q}(\omega) + \underline{Q}(\omega)\underline{h}(\Omega) = \{\underline{1} + \underline{Q}^*(-\omega)\}^{-1} - \underline{1} = \underline{F}(\Omega) \tag{4.23}$$

and its one-dimensional inversion becomes

$$H_{ab}(r) + Q_{ab}(r) + \sum_i \rho_i \int_{-\infty}^{\infty} Q_{ai}(r')H_{ib}(r - r')\mathrm{d}r' = F_{ab}(r) \qquad (4.24)$$

where

$$H_{ab}(r) = (\rho_a\rho_b)^{-1/2} \int_{-\infty}^{\infty} e^{i\omega r}h_{ab}(\omega)\mathrm{d}\omega \qquad (4.25)$$

On the other hand, Eq. (4.23) gives

$$\underline{F}(\Omega) + \underline{Q}^*(-\omega)\underline{F}(\Omega) + \underline{Q}^*(-\omega) = \underline{0} \qquad (4.26)$$

and hence we obtain its one-dimensional inversion as follows:

$$F_{ab}(r) + \sum_i \rho_i \int_{-\infty}^{\infty} Q_{ia}(-r')F_{ib}(r - r')\mathrm{d}r' + Q_{ba}(-r) = 0 \qquad (4.27)$$

The total correlation function $h_{ab}(r)$ and the direct correlation function $c_{ab}(r)$ are separated from each other by introducing the unknown function $Q_{ab}(r)$ to give Eqs. (4.24) and (4.27). There is some arbitrariness in $Q_{ab}(r)$, so that the following condition is chosen for simplicity of the integration.

$$\begin{aligned} Q_{ab}(r) &= 0: \quad r < S_{ab} = (D_a - D_b)/2 \\ Q_{ba}(-r) &= 0: \quad r > S_{ab} \end{aligned} \qquad (4.28)$$

Hence, Eq. (4.24) becomes for the integral range of $r' = S_{ai} \sim \infty$

$$H_{ab}(r) + Q_{ab}(r) + \sum_i \rho_i \int Q_{ai}(r')H_{ib}(r - r')\mathrm{d}r' = F_{ab}(r) \qquad (4.29)$$

and for the range $r' = 0 \sim S_{ai}$

$$F_{ab}(r) + \sum_i \rho_i \int Q_{ia}(-r')F_{ib}(r - r')\mathrm{d}r' + Q_{ba}(-r) = 0 \qquad (4.30)$$

The following relations hold between one- and three-dimensional Fourier transforms in the isotropic system under consideration.

$$H_{ab}(r) = 2\pi \int_r^\infty h_{ab}(r) r \, dr$$
$$C_{ab}(r) = 2\pi \int_r^\infty c_{ab}(r) r \, dr$$
(4.31)

where

$$h_{ab}(r) = -1 : \quad r < D_{ab} = (D_a + D_b)/2$$

The Percus–Yevick approximation is adopted as follows:

$$c_{ab}(r) = 0 : r \geq D_{ab}$$
$$\therefore C_{ab}(r) = 0$$

and from Eq. (4.20a)

$$Q_{ab}(r) = 0$$
(4.32)

The arbitrary function $Q_{ab}(r)$ is still unknown. However, in the case of the hard-sphere system, the range of $r > S_{ai}$ is important so that we obtain $F_{ab}(r) = 0$ from Eqs. (4.28) and (4.30), and Eq. (4.29) becomes

$$H_{ab}(r) + Q_{ab}(r) + \sum_i \rho_i \int Q_{ai}(r') H_{ib}(r - r') dr' = 0$$
(4.33)

where the integration is from $S_{ai} = (D_a - D_i)/2$ to $D_{ai} = (D_a + D_i)/2$.
In the case of equal particles, denoting the subscripts by $a = b = i = 1$

$$H_{11}(r) + Q_{11}(r) + \rho_1 \int Q_{11}(r') H_{11}(r - r') dr' = 0$$
(4.34)

where the integral range is $r' = 0 \sim D_1$. Equation (4.34) is the same as Eq. (4.8) obtained in Section 4.3.

4.6   Radial Distribution Function of Binary-Sized Particle System and Applications

Consider the random dispersion of binary-sized system of particle 1 and particle 2. The diameters and the particle number densities are, respectively, denoted by D_1, D_2 and ρ_1, ρ_2.

Placing a sphere 2 at the origin, we consider the number density of sphere 1, $\rho_1 g_{12}(r)$, at the radial distance r.

In Eq. (4.33), the subscripts are denoted by $a = 1$, $b = 2$, and there are two cases of $i = 1$ and 2. Hence,

$$H_{12}(r) + Q_{12}(r) + \rho_1 \int Q_{11}(r')H_{12}(r - r')\mathrm{d}r' + \rho_2 \int Q_{12}(r')H_{22}(r - r')\mathrm{d}r' = 0$$

$$(4.35)$$

where the Percus–Yevick approximation gives the following integral range:

the first integral is from $S_{11} = 0$ to $D_{11} = D_1$

the second integral is from $S_{12} = (D_1 - D_2)/2$ to $D_{12} = (D_1 + D_2)/2$

On the other hand, the substitution of $a = 1$ and $b = 1$ gives

$$H_{11}(r) + Q_{11}(r) + \rho_1 \int Q_{11}(r')H_{11}(r - r')\mathrm{d}r' + \rho_2 \int Q_{12}(r')H_{21}(r - r')\mathrm{d}r' = 0$$

$$(4.36)$$

where the integral range is the same as in Eq. (4.35).

The radial distribution functions $g_{12}(r)$, $g_{11}(r)$, and $g_{22}(r)$ are obtainable by using the Laplace transform method similarly to the equal particle system explained in the previous section. The solution in the Laplace transform becomes (Lebowitz, 1964; Leonard et al., 1971)

$$L\{rg_{11}(r)\} = p\{h - L_2(p)\exp(pD_2)\}/\{12(\eta_1\eta_2)^{1/2}E(p)\} \tag{4.37}$$

$$L\{rg_{12}(r)\} = p^2 \cdot \exp(pD_{12}) \cdot [\{(3/4)(\eta_2 D_2^3 - \eta_1 D_1^3)(D_2 - D_1)$$
$$- D_{12}(1 + \phi/2)\}p - (1 + 2\phi)]/E(p) \tag{4.38}$$

where $L\{\ \}$ denotes the Laplace transform and

$$h = 36\eta_1\eta_2(D_2 - D_1)^2$$

$$E(p) = h - L_1(p) \cdot \exp(pD_1) - L_2(p) \cdot \exp(pD_2) + S(p) \cdot \exp\{p(D_1 + D_2)\}$$

$$L_1(p) = 12\eta_2\{(1 + \phi/2) + (3/2)\eta_1 D_1^2(D_2 - D_1)\}D_2 p^2 + \{12\eta_2(1 + 2\phi) - hD_1\}p + h$$

$$S(p) = h + \{12(\eta_1 + \eta_2)(1 + 2\phi) - h(D_1 + D_2)\}p - 18(\eta_1 D_1^2 + \eta_2 D_2^2)p^2$$

$$- 6(\eta_1 D_1^2 + \eta_2 D_2^2)(1 - \phi)p^3 - (1 - \phi)^2 p^4$$

$$\eta_1 = \pi\rho_1/6$$

$$\eta_2 = \pi\rho_2/6$$

$$\phi = \eta_1 D_1^3 + \eta_2 D_2^3 : \text{bulk-mean particle volume fraction}$$

where ρ_1, ρ_2 and D_1, D_2 are the number density and the particle diameter, respectively. The subscripts are denoted for particle 1 and particle 2, and $D_{12} = (D_1 + D_2)/2$. Because $g_{12}(r) = g_{21}(r)$, $L\{rg_{21}(r)\}$ is obtainable from $L\{rg_{12}(r)\}$ by exchanging the subscripts 1 and 2. $L_2(p)$ is also obtainable from $L_1(p)$ by the exchange of the subscript.

The radial distribution functions $g_{12}(r)$, $g_{11}(r)$, and $g_{22}(r)$ are obtainable from the Laplace inversion of the above expressions. For further detail, see the report by Leonard et al. (1971). We now proceed to the application of the result.

4.6.1   Influence by the Presence of a Vessel Wall

Consider the random system consisting of many particle 1 and only one particle 2. Placing the particle 2 at the origin, we derive the radial distribution function of particle 1, $g_{12}(r)$, at the radial distance r.

Then the diameter of the particle 2 is made infinitive so that the particle 2 becomes equivalent to the presence of a flat plate or a wall.

Because there is only one particle 2, its number density can be set at $\rho_2 = 0$ in Eqs. (4.35) and (4.36) to become

$$H_{12}(r) + Q_{12}(r) + \rho_1 \int Q_{11}(r')H_{12}(r - r')\mathrm{d}r' = 0 \qquad (4.39)$$

$$H_{11}(r) + Q_{11}(r) + \rho_1 \int Q_{11}(r')H_{11}(r - r')dr = 0 \tag{4.40}$$

where the integral range is $r' = 0 \sim D_1$. Denoting $Q_{ab}(r) = -2\pi q_{ab}(r)$ and differentiating with respect of r,

$$rh_{12}(r) + dq_{12}(r)/dr - 2\pi\rho_1 \int q_{11}(r')(r - r')h_{12}(r - r')dr' = 0 \tag{4.41}$$

$$rh_{11}(r) + dq_{11}(r)/dr - 2\pi\rho_1 \int q_{11}(r')(r - r')h_{11}(r - r')dr' = 0 \tag{4.42}$$

$h_{11}(r) = -1$ for $r < D_1$ and hence,

$$dq_{11}(r)/dr = A_1 r + B_1 \tag{4.43}$$

$$A_1 = 1 - 2\pi\rho_1 \int q_{11}(r')dr'$$

$$B_1 = 2\pi\rho_1 \int r'q_{11}(r')dr'$$

By use of the boundary condition $q_{11}(D_1) = 0$,

$$q_{11}(r) = A_1(r^2 - D_1^2)/2 + B_1(r - D_1) \tag{4.44}$$

where

$$A_1 = (1 + 2\phi)/(1 - \phi)^2, \quad B_1 = -3D_1\phi/\{2(1 - \phi)^2\}, \quad \phi = \pi\rho_1 D_1^3/6$$

On the other hand, $h_{12}(r) = -1$ for $r < D_{12}$ and therefore

$$dq_{12}(r)/dr = A_1 r + B_1 = dq_{11}(r)/dr$$

From Eq. (4.41), we obtain

$$r\{g_{12}(r) - 1\} + A_1 r + B_1 - 2\pi\rho_1 \int q_{11}(r')(r - r')\{g_{12}(r - r') - 1\}dr' = 0$$

where the integral range is $r' = 0 \sim D_1$. A_1, B_1, and $q_{11}(r)$ are given in Eqs. (4.43) and (4.44). The Laplace transform of the above expression with respect to r, the integral of the last term, and the rearrangement gives

$$
\begin{aligned}
L\{rg_{12}(r)\} &= p^2 \exp\{p(D_1 - D_2)/2\}/\{L_2(p) - S(p)\exp(pD_1)\} \\
&\quad \times \{(D_1 + D_2 - D_1\phi + 2D_2\phi)p/2 + 1 + 2\phi\} \\
&= \{-p^2 \exp\{-p(D_1 + D_2)/2\}/S(p)\} \\
&\quad \times \sum_{n=0}^{\infty}\{L_2(p)/S(p)\}^n \exp(-npD_1) \\
&\quad \times \{(D_1 + D_2 - D_1\phi + 2D_2\phi)p/2 + 1 + 2\phi\}
\end{aligned}
\tag{4.45}
$$

where $L\{\ \}$ is the Laplace transform and p is the parameter. This is the same as Eq. (4.38) for $\eta_2 = 0$ and

$$
L_2(p) = 12\phi(1 + \phi/2)p^2/D_1^2 + 12\phi(1 + 2\phi)p/D_1^3
$$

$$
S(p) = 12\phi(1 + 2\phi)p/D_1^3 - 18\phi^2 p^2/D_1^2 - 6\phi(1 - \phi)p^3/D_1 - (1 - \phi)^2 p^4
$$

$$
\phi = \pi\rho_1 D_1^3/6 : \quad \text{bulk-mean volume fraction of particle 1}
$$

After obtaining $g_{12}(r)$ from the inversion of the Laplace transform, Eq. (4.45), D_2 is made infinitive where $(r - D_2/2)$ is kept unchanged because it is the distance from the surface of the particle 2 to the center of the particle 1. In this fashion, the radial distribution function at the dimensionless distance $x_* = r/D$ from the flat wall is derived as follows (Gotoh et al., 1978):

$$
\begin{aligned}
g(x_*) &= g^0(x_*) : \quad 0.5 \leq x_* \leq 1.5 \\
&= g^0(x_*) + g^1(x_*) : \quad 1.5 \leq x_* \leq 2.5 \\
&= g^0(x_*) + g^1(x_*) + g^{11}(x_*) : \quad 2.5 \leq x_* \leq 3.5
\end{aligned}
\tag{4.46}
$$

where

$$
g^0(x_*) = [(1 + 2\phi)/(1 - \phi)^2] \sum_{i,j,k} p_i^2 \exp\{p_i(x_* - 0.5)\}/\{(p_i - p_j)(p_i - p_k)\}
$$

$$
\begin{aligned}
g^1(x_*) &= [-24\phi(\phi + 0.5)/(1 - \phi)^4] \\
&\quad \times \sum_{i,j,k}[\{p_i V(p_i)\exp\{p_i(x_* - 1.5)\}/\{(p_i - p_j)^2(p_i - p_k)^2\}] \\
&\quad \times \{2 + W(p_i) + p_i(x_* - 1.5) - 2p_i/(p_i - p_j) - 2p_i/(p_i - p_k)\}
\end{aligned}
$$

where

$$U(p_i) = (1 + \phi/2)p_i, \quad V(p_i) = U(p_i) + 1 + 2\phi, \quad W(p_i) = U(p_i)/V(p_i)$$

$$g^{11}(x_*) = 288 \sum_{ijk} T(i,j,k)\{Q(i,j,k) - 1\}$$

$$+ 288 \sum_{ijk} T(i,j,k)\{Q(i,j,k) - W(p_i)\}W(p_i)$$

$$+ 144 \sum_{ijk} T(i,j,k)Q(i,j,k)p_i(x_* - 2.5)$$

$$- 432 \sum_{ijk} T(i,j,k)\{Q(i,j,k) - p_i/(p_i - p_j)\}p_i/(p_i - p_j)$$

$$- 432 \sum_{ijk} T(i,j,k)\{Q(i,j,k) - p_i/(p_i - p_k)\}p_i/(p_i - p_k)$$

where

$$T(i,j,k) = [\phi^2(\phi + 1/2)V(p_i)^2 \exp\{p_i(x_* - 2.5)\}]/\{(1 - \phi)^6(p_i - p_j)^3(p_i - p_k)^3\}$$

$$Q(i,j,k) = 2 + 2W(p_i) + p_i(x_* - 2.5) - 3p_i/(p_i - p_j) - 3p_i/(p_i - p_k)$$

Σ_{ijk} denotes the summation for three sets of $i, j, k = 1, 2, 3; 2, 3, 1;$ and $3, 1, 2,$ and p_i, p_j, p_k are the roots of the following third-order equation (Figure 4.1).

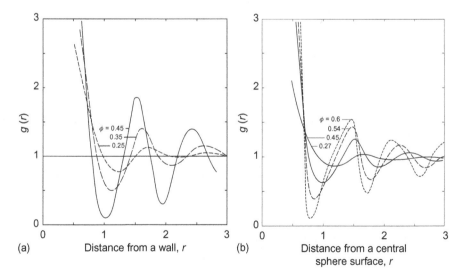

Figure 4.1 Radial distribution functions: (a) with and (b) without wall effect.

$$p^3 + \alpha_* p^2 + \beta_* p + \gamma_* = 0$$

$$\alpha_* = 6\phi/(1 - \phi), \quad \beta_* = 18\phi^2/(1 - \phi)^2, \quad \gamma_* = -12\phi(1 + 2\phi)/(1 - \phi)^2$$

4.6.2 Pore Size Distribution in Random Assemblies of Equal Spheres

Consider the spherical volume of radius R at an arbitrary point in a random assembly of equal spheres. If the volume is empty and contains no particle center, it is called the pore of radius R. The probability of existence of the pore, $P_0(R)$, in the random assembly is discussed below. The pore volume corresponds to what we call the free volume, while the spherical volume of radius $(R - D/2)$ is called the cavity, where D is the particle diameter.

$f_0(r)dr$ is defined as the probability that the spherical volume of radius r at arbitrary point 0 does not contain any particle center and the spherical shell of thickness dr contains at least one particle center. Hence, we obtain

$$P_0(R) = 1 - \int_0^R f_0(r)dr \tag{4.47}$$

$f_0(r)$ is called the Hertz distribution function. $f_0(r)dr$ can be divided into two conditional probabilities:

$$f_0(r)dr = (1 - \phi)f_1(r)dr + \phi f_2(r)dr \tag{4.48}$$

where ϕ is the bulk-mean particle volume fraction, $f_1(r)dr$ is the probability of finding the pore of radius r whose center lies outside the particles, and $f_2(r)$ dr is the probability of finding the pore of radius r whose center lies inside a particle.

For $r < D/2$, $f_1(r) = 0$ and $f_2(r)$ dr is equal to the probability of finding at least one particle center within the spherical shell of thickness dr whose center lies inside a particle. Hence,

$$f_0(r)dr = \phi 4\pi r^2 \, dr/\{\pi D^3/6\} = 24\phi r^2 \, dr/D^3 \tag{4.49}$$

Substitution into Eq. (4.47) leads to

$$P_0(R) = 1 - 4\pi R^3 \rho/3 = 1 - 8\phi(R/D)^3 : \quad R/D < 0.5 \tag{4.50}$$

In the case of $r \geqq D/2$, $f_2(r) = 0$ and

$$f_1(r)dr = \left\{ 1 - \int_{D/2}^{r} f_1(x)dx \right\} 4\pi r^2 \, dr\rho G(r) \tag{4.51}$$

The { } term on the right-hand side expresses the probability of finding no particle centers in the spherical volume of radius r, the second term is the probability of finding at least one particle center in the spherical shell of radius r and thickness dr, ρ is the number density, and $G(r)$ is the radial distribution function at the pore surface. Solving Eq. (4.51) for $f_1(r)$, it is substituted into Eq. (4.48) and from Eq. (4.47), one obtains

$$P_0(R) = (1 - \phi)\exp\left\{ -24\phi \int_{D/2}^{R} r^2 G(r)dr/D^3 \right\} : \quad R/D \geqq 0.5 \tag{4.52}$$

The radial distribution function $g(r)$ is the magnification factor for the bulk-mean value to obtain the local particle number concentration at distance r from an arbitrary particle center. Here, we need its value on the pore surface.

Consider a binary-sized particle system. The radial distribution function of sphere 1 about sphere 2, $g_{12}(r)$, was obtained by the Percus–Yevick approximation in the previous section. Particle 2 is supposed to be only one and then $n_2 = 0$. If particle 2 is regarded as the pore and by setting $r = (D_1 + D_2)/2$, we obtain $g_{12}(r) = G(r)$:

$$G(r) = \{(1 + 2\phi) - 3\phi/(2r/D)\}/(1 - \phi)^2 \tag{4.53}$$

And from Eq. (4.52),

$$P_0(R) = (1 - \phi)\exp[\phi(1 - \phi)^{-2}\{- 8(1 + 2\phi)(R/D)^3 + 18\phi(R/D)^2 + 1 - 5\phi/2\}] \tag{4.54}$$

where $R/D \geqq 0.5$ (Gotoh et al., 1986).

4.6.3   Size Distribution of Aggregates Inherent in Random Dispersion of Equal Spheres

The spherical volume of radius r is considered at an arbitrary particle center. $f(r)dr$ is defined as the probability that the spherical volume does not contain any other particle center and the spherical shell of thickness dr contains at least one particle center. Hence, we obtain

$$f(r)\mathrm{d}r = \left[1 - \int_1^r f(x)\mathrm{d}x\right] 4\pi r^2 \, \mathrm{d}r \rho g(r) \tag{4.55}$$

where the distance r is dimensionless in unit of the particle diameter D. The [ ] term in the right-hand side is the probability that the spherical volume of radius r does not contain any particle centers other than the central one. Equation (4.55) yields

$$f(r) = - \mathrm{d}/\mathrm{d}r\left[\exp\left\{-24\phi\int_1^r x^2 g(x)\mathrm{d}x\right\}\right] \tag{4.56}$$

where $\phi = \pi D^3 \rho/6$: bulk-mean particle volume fraction.

The average nearest neighbor distance from the central particle becomes in unit of the particle diameter as follows:

$$\langle R \rangle = \int_1^\infty r f(r)\mathrm{d}r = 1 + \int_1^\infty \exp\left\{-24\phi\int_1^r x^2 g(x)\mathrm{d}x\right\}\mathrm{d}r \tag{4.57}$$

$y(-)$	$\bar{R}(-)$
0.03	1.616
0.05	1.432
0.08	1.297
0.10	1.243
0.12	1.204
0.15	1.160
0.17	1.139
0.20	1.113
0.22	1.0989
0.25	1.0818
0.27	1.0723
0.30	1.0602
0.35	1.0446
0.40	1.0331

The above table shows the $\langle R \rangle$ versus ϕ relation obtained from the radial distribution function by the Percus–Yevick approximation, where $y = \phi$ and $\bar{R}$ denotes $\langle R \rangle$.

Next the size distribution of aggregates inherent in the random dispersion of equal spheres is considered by assuming that any two particles within the critical distance Rc are regarded as being connected with each other, where Rc is the aggregation radius in unit of the particle diameter.

The probability that two neighboring particles are connected with each other becomes

$$P = \int_1^{Rc} f(x)dx = 1 - \exp\left[-24\phi \int_1^{Rc} x^2 g(x)dx\right] \tag{4.58}$$

$g(x)$ by the Percus−Yevick approximation gives the P versus Rc relation as depicted in the below figure.

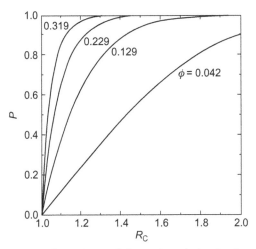

Next consider the size of aggregates inherently existing in the random dispersion of equal spheres. The aggregate size Nc means that it consists of Nc particles. $Q(Nc)$ denotes the probability of finding the aggregate of size Nc.

Two neighboring particles denoted by a and b are supposed to be connected each other with the probability P. The particle a is isolated and the particle b belongs to an aggregate of size Nc. Hence, the probability that the aggregate of size (Nc + 1) is made by connecting a and b is expressed by

$$Q(Nc + 1) = P \cdot Q(Nc) \tag{4.59}$$

where Nc = 1,2,... and $Q(1) = 1 - P$ is the probability that the particle remains isolated. Therefore,

$$Q(Nc) = (1 - P)P^{Nc-1} \tag{4.60}$$

$Q(1) = 1 - P = n_1/N_t$ is depicted in the below figure comparing with the computer experiments, where n_1 is the number of isolated particles, and N_t is the total number of particles.

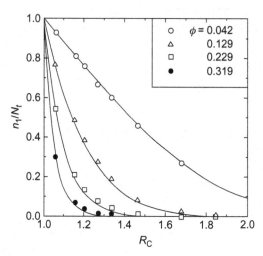

The average size of aggregates, that is, the average number of the particles contained in a single aggregate, becomes as follows:

$$N_{50} = -\ln 2/\ln P \qquad\qquad (4.61)$$

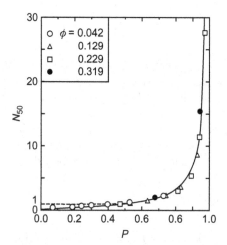

Equation (4.61) is depicted in the above figure comparing in good agreement with the computer experiments.

In summary, there exists inherently a wide distribution in the aggregate sizes, even in the random dispersion.

References

Baxter, R.J., Ornstein–Zernike relation and Percus–Yevick approximation for fluid mixtures, J. Chem. Phys., vol. 52, 4559 (1970).

Gotoh, K., Jodrey, W.S., and Tory, E.M., Variation in the local packing density near the wall of a randomly packed bed of equal spheres, Powder Technol., vol. 20, 257 (1978).

Gotoh, K., Nakagawa, M., and Matsuoka, H., Size distribution of clusters inherent in random dispersions of equal spheres, Part. Sci. Technol., vol. 3, 27 (1985).

Gotoh, K., Nakagawa, M., Furuuchi, M., and Yoshigi, A., Pore size distribution in random assemblies of equal spheres, J. Chem. Phys., vol. 85, 3098 (1986).

Lebowitz, J.L., Exact solution of generalized Percus–Yevick equation for a mixture of hard spheres, Phys. Rev., vol. 133, A895 (1964).

Leonard, P.J., Henderson, D., and Barker, J.A., Calculation of the radial distribution function of hard-sphere mixtures in the Percus–Yevick approximation, Mol. Phys., vol. 21, 107 (1971).

Mandel, F., Bearman, R.J., and Bearman, M.Y., Numerical solutions of the Percus–Yevick equation for the Lennard-Jones (6,12) and hard sphere potentials, J. Chem. Phys., vol. 52, 3315 (1970).

Smith, W.R., and Henderson, D., Analytical representation of the Percus–Yevick hard-sphere radial distribution function, Mol. Phys., vol. 19, 411 (1970).

Throop, G.J., and Bearman, R.J., Numerical solutions of the Percus–Yevick equation for the hard sphere potential, J. Chem. Phys., vol. 42, 2408 (1965).

5 Sample Size for Measuring Particle Concentration

5.1 Distribution Functions

This section outlines the relationship of functions among the Bernoulli distribution, the Poisson distribution, and the Gaussian distribution.

N equal particles are supposed to be distributed randomly in the large volume V. The probability of finding a particle in the small volume v becomes v/V. There are $N!/[n!(N-n)!]$ cases for choosing n particles out of N. Hence, the probability of finding n particles in the small volume v becomes

$$W_N(n) = \{N!/[n!(N-n)!]\}(v/V)^n\{(V-v)/V\}^{N-n} = \{N!/[n!(N-n)!]\}p^n q^{N-n}$$

This is the Bernoulli distribution, where $p = v/V$, $q = 1 - p$, and

$$\sum_{n=0}^{N} W_N(n) = [(pu+q)^N]_{u=1} = 1$$

The average of n is

$$\langle n \rangle = \sum_{n=0}^{N} n W_N(n)$$

$$= \sum_{n=0}^{N} n \,[\text{the coefficient of } u^n \text{ in series expansion of } (pu+q)^N]$$

$$= \sum_{n=0}^{N} [\text{the coefficient of } u^n \text{ in series expansion of } \mathrm{d}/\mathrm{d}u(pu+q)^N]$$

$$= [\mathrm{d}/\mathrm{d}u(pu+q)^N]_{u=1}$$

$$= Np(p+q)^{N-1} = Np = Nv/V = \nu$$

Particulate Morphology. DOI: 10.1016/B978-0-12-396974-3.00005-9

Using the average number $\nu = \langle n \rangle$,

$$W_N(n) = \{N!/[n!(N-n)!]\}(\nu/N)^n(1-\nu/N)^{N-n}$$
$$= (\nu^n/n!)\cdot 1\cdot (1-1/N)\ (1-2/N)\ \cdots\ \{1-(n-1)/N\}\ (1-\nu/N)^{N-n}$$

In many applications, we can set $N \to \infty$ and $V \to \infty$. Hence,

$$W(n) = \lim_{N\to\infty} W_N(n) = (\nu^n/n!)\lim_{N\to\infty}(1-\nu/N)^N$$
$$= \nu^n\, e^{-\nu}/n!$$

This is the Poisson distribution. ν is large in practical applications so that $W(n)$ becomes very small, except the region of $n \doteq \nu$. In this case, the Poisson distribution can be approximated further to become the Gaussian distribution as shown below:

$$\ln W(n) = n \ln \nu - \nu - \ln n!$$

Applying the Stirling formula to the factorial term,

$$\ln W(n) = n \ln \nu - \nu - \ln [(2\pi\nu)^{1/2}n^n\, e^{-n}]$$

and substituting $n = \nu + \delta = \nu(1 + \delta/\nu)$ leads to

$$\ln W(n) = \delta - (\ln 2\pi\nu)/2 - (\nu + \delta + 1/2)\ln(1 + \delta/\nu)$$

In the case of $\delta/\nu \doteq 0$, $\nu + \delta + 1/2 \doteq \nu + \delta$ and

$$\ln (1 + \delta/\nu) \doteq \delta/\nu - (\delta/\nu)^2/2$$

Therefore,

$$\ln W(n) = -\delta^2/(2\nu) + \delta^3/(2\nu^2) - (\ln 2\pi\nu)/2$$
$$\doteq -\delta^2/(2\nu) - (\ln 2\pi\nu)/2$$

and we obtain

$$W(n) = (2\pi\nu)^{-1/2} \exp[-\delta^2/(2\nu)]$$
$$= (2\pi\nu)^{-1/2} \exp[-(n - \nu)^2/(2\nu)]$$

This is the Gaussian distribution, i.e., the normal distribution.

5.2 Sample Size of Measurement

Consider the concentration measurement of particulate materials such as pollen and exhaust (black smoke) suspended in the air. Because the air is too vast to treat mathematically, we consider the concentration measurement of the particulate materials in the room of volume V. If all the amount of the particulate materials is collected, the bulk-mean concentration is obtainable by dividing it by V. It is, however, impossible for us to collect all the materials being suspended in the room. Accordingly, we should consider a local measurement in the small space of volume v, leading to the problem of the sampling size (Gotoh, 1971).

Space volume of a randomly dispersed system of equal-sized particles is supposed to consist of N cubic cells, each size of which is equal to the particle diameter. If the dispersed system contains M particles, M of the N cells are filled with particles and the ratio M/N relates to the bulk-mean number density of particles.

In the case of complete randomness, the number of possible distributions of M particles among N cells is

$$T = N!/[M!(N - M)!]$$

Similarly, the local space volume of a sample is supposed to consist of n cells and contain m particles. The local mean number density is given by m/n. The number of cases in which the local space volume contains m particles is expressed by

$$T_m = n!/[m!(n - m)!] \times (N - n)!/[(M - m)!\{(N - n) - (M - m)\}!]$$

because m particles are shared among n cells, and the remaining $(M - m)$ particles are shared among remaining $(N - n)$ cells. If each particle distribution is equally probable, i.e., in the case of complete randomness, the local space volume of the sample shows the number density $C_m = m/n$ with the following probability:

$$P(C_m) = T_m/T$$
$$= [2\pi NRC(1 - C)(1 - R)]^{-1/2} \exp[-NR(C_m - C)^2/\{2C(1-C)(1-R)\}]$$

where the factorial terms are approximated by the Stirling series and $R = n/N$. Therefore, the confidence coefficient for the confidence interval $C(1 - e) \sim C(1 + e)$ is

$$P = \sum_{C_m=C(1-e)}^{C(1+e)} P(C_m) = n \int_{C(1-e)}^{C(1+e)} P(C_m)dC_m$$
$$= (2/\pi)^{1/2} \int_0^{up} \exp(-u^2/2)du$$

where

$$up = Ce/\sigma, \quad (C_m - C)/\sigma = u, \quad \sigma^2 = C(1 - C)(1 - R)/(NR)$$

In other words, the local space volume of the sample shows the bulk-mean number density C with probability P and within the allowable error e.
$P = 0.99$ for $up = 2.5758$ and hence,

$$m = (1 - C)(2.5758/e)^2 : \quad n \ll N (= \text{total volume})$$

With 99% confidence, we can say that the sample measured by the size n shows the bulk-mean number density C within the error e.
For example,

1. $m = 1036(1 - C)$ for $e = 0.08$
2. $m = 2653(1 - C)$ for $e = 0.05$
3. $m = 10616(1 - C)$ for $e = 0.025$

The concentration of particulate materials is very low in the air and hence $C \doteqdot 0$ in general, yielding $m = 1036$ for (1). In other words, the sampling volume should contain $m = 1036$ particles in order to measure the bulk-mean number density of the particles within 8% error and with 99% confidence. Needless to say, the larger the sample volume, the more accurate the measurement of the bulk-mean value.

Next, consider the tally of election returns as an example of large C. The unit cell under consideration corresponds to a single vote and N is the total number of the votes. If a candidate gains M votes and $C = M/N = 0.51$, he or she will win by a narrow majority. When can one predict the win? From (1), we obtain $m = 1036$

$(1-0.51) = 508$ and hence, $n = m/C = 996$. Namely, if 996 votes are taken out randomly and 508 votes are for the particular candidate, it can be said with 99% confidence that the candidate will win within 8% error. In the case of an easy win, the sample size required for the prediction becomes much smaller than 1000.

In summary, for measuring the bulk-mean number density, the sample size can be determined from (1) for 8% error and from (2) for 5% error.

Reference

Gotoh, K., Spatial distribution of equal-sized particles in a random assemblage and minimum size of a representative sample, Ind. Eng. Chem. Fundam., vol. 10, no. 1, 161 (1971).

6 Introduction to Statistical Thermodynamics

6.1 Quantum States of Steady Thermal Vibration

A single particle is supposed to be in the steady state of thermal vibration, where we consider the possible number of quantum states. In quantum mechanics, the particle motion in one dimension is treated as being equivalent to the wave vibration. One-dimensional steady waves confined in the interval L have the wavelength $\lambda = 2L/n$, where $n = 1, 2, \ldots$. The wave velocity c and the frequency ν have the relation $\lambda\nu = c$, and hence,

$$c/\lambda = cn/2L = \nu_n$$

where ν_n is the frequency of steady vibration for $n = 1, 2, \ldots$.

On the other hand, there exists the de Broglie relation between the wavelength λ and the momentum p of particle motion:

$$p = h_p/\lambda$$

where h_p is the Planck constant. The kinematic energy ε is expressed by $\varepsilon = p^2/(2m)$ and

$$\varepsilon = h_p^2/(2m\lambda^2)$$

Substitution of $\lambda = 2L/n$ leads to

$$\varepsilon = [h_p^2/(8mL^2)]n^2$$

Particulate Morphology. DOI: 10.1016/B978-0-12-396974-3.00006-0

This is the intrinsic energy for the quantum state of one-dimensional thermal vibration with the quantum number n.

Next, consider the steady two-dimensional vibration of the thin film that is a square with fixed peripheries. One side of the square is taken as x-axis and the moving direction of the steady wave is denoted by θ from x-axis. Hence, the wavelength λ can be expressed by x- and y-components:

$$\lambda_x = \lambda/\cos\theta, \quad \lambda_y = \lambda/\sin\theta$$

In order to be steady, the wavelengths must obey the following conditions:

$$\lambda_x = 2L/n_x, \quad \lambda_y = 2L/n_y$$

where L is the side length of the square and n is the quantum number with each subscript denoting the wave component.

Using $\sin^2\theta + \cos^2\theta = 1$,

$$(\lambda/\lambda_x)^2 + (\lambda/\lambda_y)^2 = 1$$

and hence,

$$[\lambda^2/(4L^2)](n_x^2 + n_y^2) = 1 \quad \therefore \lambda = 2L/(n_x^2 + n_y^2)^{1/2}$$

This is the wavelength of steady vibration for the set of quantum number (n_x, n_y). The frequency ν of steady transverse wave becomes as follows:

$$\nu = c/\lambda = [c/(2L)](n_x^2 + n_y^2)^{1/2}$$

where c is the wave velocity.

Similarly, the wavelength of three-dimensional steady vibration in the cube of size L becomes as follows:

$$\lambda = 2L/(n_x^2 + n_y^2 + n_z^2)^{1/2}$$

and the frequency is expressed by

$$\nu = c/\lambda = [c/2L](n_x^2 + n_y^2 + n_z^2)^{1/2}$$

The set of (n_x,n_y,n_z) gives the intrinsic frequency, where the point of $n_x = n_y = n_z = 0$ is excluded.

Consider the rectangular coordinates of (n_x,n_y,n_z). Each lattice point corresponds to the quantum state for steady thermal vibration. The region under consideration is one-eighth of the total space, because $n_x > 0$, $n_y > 0$, and $n_z > 0$. The spherical volume $(4\pi r^3/3)/8$ contains approximately the same number of lattice points in the case of large radius r. Accordingly,

$$(4\pi r^3/3)/8 = (\pi/6)(2L/\lambda)^3 = (4\pi/3)V(2m\varepsilon)^{3/2}/h_p^3$$

where $\lambda = 2L/r$, $p\lambda = h_p$, $p^2 = 2m\varepsilon$, and $L^3 = V$ (volume). This is the total number of the quantum states for a single particle with the energy smaller than ε.

6.2   Analytical Dynamics and Generalized Coordinates

In classical dynamics, the motion of a free particle is expressed by

$$(p_x^2 + p_y^2 + p_z^2) = 2m\varepsilon$$

where m is the mass, ε is the energy, and p is the momentum with each subscript denoting x-, y-, and z-components. Accordingly, the region of the energy smaller than ε in the generalized coordinates is

$$0 < (p_x^2 + p_y^2 + p_z^2) < (2m\varepsilon)^{1/2}$$

and therefore its volume becomes as follows:

$$(4\pi/3)(2m\varepsilon)^{3/2}V$$

where $V = \iiint dq$, $dq = dx\,dy\,dz$.

V is the volume under consideration; p and q are called the canonical variables.

Comparing the above volume with that in the quantum coordinates in Section 6.1, we can say that the volume in the generalized coordinates divided by the minimum volume h_p^3 yields the total number of quantum states for a free particle.

6.3   Stationary Distribution and Partition Function

n_i particles with energy ε_i are supposed to be in the quantum state:

$$\sum n_i = N, \quad \sum \varepsilon_i n_i = E$$

where N is the total number of particles and E is the total energy.

Under the constants N and E, the most probable distribution of particles in the energy states is considered below.

The total number M of the partitioning is

$$M = N!/(n_1! n_2! n_3! \cdots)$$

Stirling formula: $\ln M = N(\ln N - 1) - \sum n_i (\ln n_i - 1)$.
The condition of extremum for M is

$$\sum \delta n_i = 0, \quad \sum \varepsilon_i \, \delta n_i = 0$$

$$\delta \ln M = 0 \quad \text{or} \quad \sum \ln n_i \, \delta n_i = 0$$

and Lagrange's method of indeterminate coefficients gives

$$\sum \ln n_i \, \delta n_i + \left(\sum \delta n_i \right) \times \alpha + \left(\sum \varepsilon_i \, \delta n_i \right) \times \beta = 0$$

$$\therefore \ln n_i = -\alpha - \beta \varepsilon_i$$

$$n_i = e^{-\alpha} \exp(-\beta \varepsilon_i) = (N/Q_1) \exp(-\beta \varepsilon_i)$$

$$n_i/N = \exp(-\beta \varepsilon_i)/Q_1$$

where Q_1 is used in place of α. The indeterminate coefficients Q_1 and β are obtainable from the following conditions:

$$Q_1 = \sum \exp(-\beta \varepsilon_i)$$

$$\sum \varepsilon_i \exp(-\beta \varepsilon_i) / \left\{ \sum \exp(-\beta \varepsilon_i) \right\} = E/N = E_{av}$$

Consider a small change of the energy:

$$dE = \sum \varepsilon_i \, dn_i + \sum n_i \, d\varepsilon_i$$

The second term of the right-hand side expresses the work done from outside of the system and is denoted by $d'W$.

$$d'W = \sum n_i \, d\varepsilon_i = N \sum d\varepsilon_i \exp(-\beta\varepsilon_i) / \left\{ \sum \exp(-\beta\varepsilon_i) \right\} = dE_{av}$$

$$d(\ln Q_1) = dQ_1 / Q_1 = -\left[\sum \varepsilon_i \exp(-\beta\varepsilon_i) / \left\{ \sum \exp(-\beta\varepsilon_i) \right\} \right] d\beta$$

$$-\beta \sum d\varepsilon_i \exp(-\beta\varepsilon_i) / \left\{ \sum \exp(-\beta\varepsilon_i) \right\}$$

$$= -E_{av} d\beta - \beta \, dE_{av} = (E_{av}/k)(dT/T^2) - dE_{av}/(kT)$$

On the other hand, using the Helmholtz free energy F ($=U - TS$) in thermodynamics,

$$dF = d'W - S \, dT$$

$$\therefore \ \ d(-F/T) = (U/T^2)dT - dE_{av}/T$$

Therefore,

$$\beta = 1/kT$$

$$F = -kT \ln Q_1, \quad Q_1 = \sum \exp(-\varepsilon_i/kT)$$

Q_1 is called the partition function that expresses the total sum of partitioning fractions for possible energy states of the single particle.
$Q = Q_1^N$ for the system of N free particles, while $Q = Q_1^N/N!$ for indistinguishable particles such as gases. Accordingly,

$$Q = Q_1^N/N! = \int \cdots \int \exp[-\varepsilon/(kT)]d^{3N}p \, d^{3N}q/(N!h_p^{3N})$$

$$= (V^N/N!)(2\pi mkT/h_p^2)^{3N/2}$$

where $\varepsilon = \sum_1^N (p_x^2 + p_y^2 + p_z^2)/(2m)$, $\int \ldots \int d^{3N} q = V^N$ and the summation is for N particles.

6.4 Number Density and Distribution Function

N molecules are supposed to be placed at r_1, r_2, ..., and r_N. The number density function is defined by

$$v^{(1)}(r) = \sum_1^N \delta(r - r_i)$$

the pair density function is defined by

$$v^{(2)}(r, r') = \sum_{i=1}^N \sum_{k=1}^N \delta(r - r_i)\delta(r' - r_k): \quad i \neq k$$

and so on.

For the total volume V, the following relations hold:

$$\int_v v^{(1)}(r) dr = N, \quad \iint_v v^{(2)}(r, r') dr\, dr' = N(N - 1)$$

In thermal equilibrium at absolute temperature T, the probability of finding molecules in the small volume dr_1 at r_1, in the small volume dr_2 at $r_2, \ldots$, and in the small volume dr_N at r_N is proportional to the following expression:

$$\exp[-\phi(r_1, r_2, \ldots, r_N)/kT] dr_1\, dr_2 \ldots dr_N$$

where $\phi(r_1, r_2, \ldots, r_N)$ is the potential energy of the total system.

The spatial average of the number density function is called the density distribution function:

$$\rho^{(1)}(r) = \langle v^{(1)}(r) \rangle$$
$$= \int \cdots \int \exp[-\phi(r_1, r_2, \ldots, r_N)/kT] \sum_1^N \delta(r - r_i) dr_1\, dr_2 \ldots dr_N /$$
$$\int \cdots \int \exp[-\phi(r_1, r_2, \ldots, r_N)/kT] dr_1\, dr_2 \ldots dr_N$$

Similarly, the pair distribution function becomes

$$\rho^{(2)}(r,r') = \langle v^{(2)}(r,r') \rangle$$

$$= \int \cdots \int \exp[-\phi(r_1,r_2,\ldots,r_N)/kT] \sum_{i=1}^{N}\sum_{k=1}^{N}\delta(r-r_i)\delta(r'-r_k)dr_1\,dr_2\ldots dr_N \Big/$$

$$\int \cdots \int \exp[-\phi(r_1,r_2,\ldots,r_N)/kT]dr_1\,dr_2\ldots dr_N$$

where $i \neq k$ and the integral is for the total space volume.

$$\int \rho^{(2)}(r,r')dr' = N(N-1)\rho^{(1)}(r)$$

For a fixed particle at r, the number density of particles at r' is denoted by $^r\rho^{(2)}(r,r')$. The number of pairing between particles in the small volume dr and those in dr' is $\rho^{(2)}(r,r')dr\,dr'$. For a fixed particle in dr, the pairing number becomes $\rho^{(2)}(r,r')dr\,dr'/\rho^{(1)}(r)dr$, dividing it by the small volume dr' yields the number density at r' as follows:

$$^r\rho^{(2)}(r,r') = \rho^{(2)}(r,r')/\rho^{(1)}(r)$$

$$= (N-1)\int \cdots \int \exp[-\phi(r_1,r_2,\ldots,r_N/kT]dr_3\ldots dr_N \Big/$$

$$\int \cdots \int \exp[-\phi(r_1,r_2,\ldots,r_N)/kT]dr_2\ldots dr_N$$

When a molecule is fixed at r, the others cannot approach within the region closer than the molecular diameter. In other words, the distribution of molecules is affected by the presence of the fixed one.

In an isopropic system, $\rho^{(2)}(r,r')$ is independent of r and becomes a function of $|r-r'|$ only. And the correlation function $g(r,r')$ is introduced as follows:

$$\rho^{(2)}(r,r') = \rho^{(1)}(r)\rho^{(1)}(r')g(r,r')$$

$g(r,r') = 1$ for a uniform system.

For gas molecules in no external force,

$$\rho^{(1)}(r) = \text{constant} = \rho = N/V$$

$$\rho^{(2)}(r, r') = \rho^2 g(R) : \quad R = |r - r'|$$

$g(R)$ is called the radial distribution function, and hence,

$${}^r\rho^{(2)}(r, r') = \rho g(R)$$

6.5   Equation of State for Gases

The Helmholtz free energy is expressed by

$$F = -kT \ln Q$$

and from the thermodynamics relation,

$$p = -(\partial F / \partial V)_T = kT(\partial \ln Q / \partial V) = kT(\partial Q / \partial V)/Q$$

the pressure becomes $p = NkT/V$ for an ideal gas.

In the case of real gases, the total energy of the system is expressed by

$$\varepsilon = \sum_{1}^{N} (p_x^2 + p_y^2 + p_z^2)/(2m) + \phi$$

where ϕ is the potential energy and

$$Q = Z/\lambda^{3N}, \quad \lambda = h_p/(2\pi mkT)^{1/2}$$

$$Z = (N!)^{-1} \int \cdots \int \exp[-\phi/(kT)] dx_1 \, dy_1 \, dz_1 \cdots dx_N \, dy_N \, dz_N$$

$$p = kT(\partial \ln Z / \partial V)_T$$

The gas molecules are supposed to be in a cube of size L and all variables are made dimensionless as follows:

$$\xi_1 = x_1/L \quad \eta_1 = y_1/L \quad \zeta_1 = z_1/L$$

$$\vdots$$

$$\xi_N = x_N/L \quad \eta_N = y_N/L \quad \zeta_N = z_N/L$$

The potential energy is assumed to be the sum for all molecular pairs and hence,

$$\phi = \sum \phi(LS_{ik}) : \quad |r_i - r_k| = R_{ik} = LS_{ik}$$

Using $V = L^3$ and $dV = 3L^2\, dL$,

$$(\partial \ln Z/\partial V)_T = 1/(3L^2)(\partial \ln Z/\partial L)$$

$$= N/V + (3L^2)^{-1} \int \cdots \int \left\{ -\sum S_{ik} \phi'(LS_{ik})/kT \right\} \exp[-\phi/(kT)]$$

$$\times d\xi_1\, d\eta_1\, d\zeta_1 \ldots d\xi_N\, d\eta_N\, d\zeta_N /$$

$$\int \cdots \int \exp[-\phi/(kT)] d\xi_1\, d\eta_1\, d\zeta_1 \cdots d\xi_N\, d\eta_N\, d\zeta_N$$

$$= N/V - (3VkT)^{-1} \int \cdots \int \sum R_{ik} \phi'(R_{ik}) \exp\left[-\sum R_{ik} \phi(R_{ik})/(kT) \right]$$

$$\times dr_1 \cdots dr_N / \int \cdots \int \exp\left[-\sum R_{ik} \phi(R_{ik})/(kT) \right] dr_1 \cdots dr_N$$

$$= N/V - N(N-1)(6VkT)^{-1} \iint R_{12} \phi'(R_{12}) dr_1\, dr_2$$

$$\times \int \cdots \int \exp[-\phi/(kT)]\, dr_3 \cdots dr_N / \int \cdots \int \exp[-\phi/(kT)] dr_1 \cdots dr_N$$

The pair distribution function is as follows:

$$^r\rho^{(2)}(r, r') = N(N-1) \int \cdots \int \exp[-\phi(r_1, r_2, \ldots, r_N)/kT] dr_3 \cdots dr_N /$$

$$\int \cdots \int \exp[-\phi(r_1, r_2, \ldots, r_N)/kT] dr_1 \cdots dr_N$$

Accordingly,

$$(\partial \ln Z / \partial V)_T = N/V - (6VkT)^{-1} \iint R_{12} \phi'(R_{12}) dr_1 \ dr_2 \ ^r\rho^{(2)}(r, r')$$

$$= N/V - \rho^2(6VkT)^{-1}V \cdot 4\pi \int_0^\infty R^3 \phi'(R)g(R)dR$$

and the equation of state for gases becomes

$$pV/(NkT) = 1 - 2\pi N(3kTV)^{-1} \int_0^\infty R^3 \phi'(R)g(R)dR$$

In the case of hard-sphere gases of diameter σ and for $\delta \ll \sigma$,

$$\phi(R) = \phi(R) \geq kT \quad : \quad R \leq \sigma - \delta$$
$$= \phi(R) \qquad\quad : \quad \sigma - \delta \leq R \leq \sigma$$
$$= 0 \qquad\qquad : \quad R > \sigma$$

$g(R) = g(\sigma)$ at $R = \sigma$ and therefore,

$$g(R) \ = 0 \qquad\qquad\qquad\qquad : \quad R \leq \sigma - \delta$$
$$= g(\sigma)\exp[-\phi(R)/kT] \quad : \quad \sigma - \delta \leq R \leq \sigma$$
$$\phi'(R) = d\phi(R)/dR = 0 \qquad : \quad R > \sigma$$

In the limit $\delta \to 0$, we obtain

$$\int_0^\infty R^3 \phi'(R)g(R)dR = g(\sigma) \int_{\sigma-\delta}^\sigma R^3 \phi'(R)\exp[-\phi(R)/kT]dR$$

$$= -kTg(\sigma)\sigma^3$$

Hence, the equation of state for hard-sphere gases becomes as follows:

$$pV/(NkT) = 1 + 2\pi N(3V)^{-1}\sigma^3 g(\sigma)$$

At the end of Section 4.4, the analytical solution of $g(\sigma)$ was given as follows:

$$g(\sigma) = (1 + \phi/2)/(1 - \phi)^2$$

Finally, the equation of state for a hard-sphere gas model becomes

$$pV/(NkT) = (1 + 2\phi + 3\phi^2)/(1 - \phi)^2$$

7 Particulate Morphology

Although structures of molecular configurations change at every moment due to thermal vibration, liquids can be regarded as random packing of particles. Since Bernal and Scott first proposed this idea, many investigations have been made into the spatial distribution of particles. The structural study can play important roles in many fields of science and engineering. The radial distribution function in statistical thermodynamics can express physicochemical properties of matter. Fluidization in chemical engineering is a structural problem. If we take physical and chemical properties away from our subject, the remainder is the structure itself. Hence, the structural study is ubiquitous and of fundamental importance.

Focusing attention on the structural study, this chapter summarizes a series of the author's research works on the comparison between packing structures of molecular systems and powder assemblies. In what follows, we first discuss the structure of a random assembly of equal spheres. We then discuss the packing situation of molecular systems through simplified statistical thermodynamics. And, finally, we compare the packing structures between molecular systems and powder assemblies.

Figure 7.1 illustrates the structural comparison of the molecular system and the random particle assemblies (Gotoh, 1971a; 1980).

It is better for us to explain final results first. Figure 7.1 illustrates a general comparison between packing structures of molecular systems and powder assemblies, which compares the two systems in terms of the bulk-mean volume fraction ϕ of the particles; λ is the molecular shape parameter. For a while, we will consider the case of $\lambda = 1$ only, i.e., the case of spherical particles. All numerical values at boundaries of the regions are obtained experimentally from both the sphere assembly and the thermodynamics of simple fluids. First, consider the molecular system. When the concentration of molecules is very low, each molecule behaves as an individual and is in the gaseous state. At the critical point, the average spacing of the neighboring particles becomes only about 10% of the particle diameter, and the molecules are considered to be in quasi-contact due to attractive forces among them. Accordingly, in concentrations higher than the critical point, the molecules behave as the gas-like liquid. The region between the boiling and the melting points corresponds to what we call the liquid state, and the molecules are in mutual

Particulate Morphology. DOI: 10.1016/B978-0-12-396974-3.00007-2

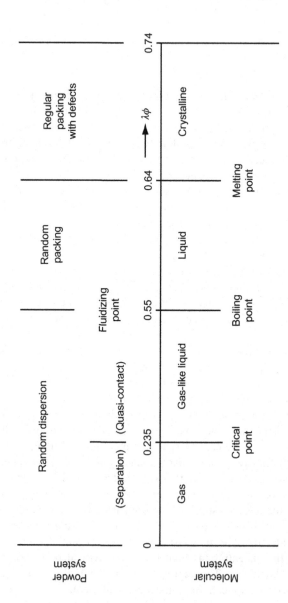

Figure 7.1 General comparison between packing structures of molecular systems and powder assemblages. λ = molecular shape parameter ($\lambda = 1$ for spheres). ϕ = bulk-mean particle volume fraction.
Source: From Gotoh (1971a) with permission.

contact. The boiling point corresponds to the random loose packing of particles; the melting point to the random close packing. In the concentration higher than the melting point, the random packing cannot exist. As a result, some regularity should appear; this means the occurrence of crystalline. In summary, Figure 7.1 shows a generalized classification of the state of matter based on the bulk-mean volume fraction of the particles. From random packing of equal spheres, we have experimental evidences that the bulk-mean volume fraction of particles is 0.6366 for the random close packing and 0.601 for the random loose packing. Computer-simulated loose packing yields $\phi = 0.582$, while the spherical powder system gently settled down after fluidization yields $\phi = 0.56$. The closest packing density of equal spheres is of course 0.74. Although this is a rough explanation, no mathematical procedure can explain all of these critical packing densities. We do not know how to express the packing situation in detail. In the concentration lower than $\phi = 0.55$, each particle is not in contact with others overall, so we can describe this region as being the particle dispersion. As mentioned before, in the range higher than $\phi = 0.235$, the particles are in quasi-contact due to attractive forces among them. The range between $\phi = 0.55$ and 0.64 corresponds to the random packing. The range higher than $\phi = 0.64$ should correspond to the regular packing. However, there are only two types of the regular packing in this region, namely, the tetragonal-sphenoidal and rhombohedral packings. Accordingly, if a continuous change in the packing density does occur, a randomness other than the previous criterion must appear in the system, such as the regular packing with randomly distributed defects. In this fashion, we have obtained the general comparison between packing structures of molecular systems and powder assemblies as shown in Figure 7.1. As mentioned before, no mathematical procedure is available for explaining the detailed structure of the particle assemblies.

7.1   Random Packing

Consider the random packing of equal spheres. Bearing balls are supposed to be packaged in a vessel. The bulk-mean particle volume fraction becomes 0.601 in the case of the loose packing, while in the tightest packing it becomes 0.6366.

In the experiment by Finney (1970), bearing balls of diameter 6 mm were packaged randomly in a vessel. The ball arrangement becomes regular in general due to the presence of the flat wall. Accordingly, some idea should be devised for the wall surface in order to keep the random arrangement of the balls. This is another subject of the research. He made the tightest packing of about 7000 balls confined in a rubber vessel with an uneven surface, into which melted wax was poured to fix the ball position. After the wax set, (x,y,z) coordinates of each ball were measured one by one. The packing structure was displayed as an expanded model using ping-pong balls and matchsticks.

Once the data of (x,y,z) coordinates are stored in the computer memory, it is easy for us to obtain the geometrical structure of the ball assembly, leading to the bulk-mean volume fraction of balls in the tightest packing: $\phi = 0.6366$ that is close to $2/\pi$ and its mathematical proof is desirous.

Next, consider the random loose packing whose bulk-mean particle volume fraction is 0.601. It becomes looser in the case of no gravity. For example, light ping-pong balls are supposed to be fluidized in a vessel by upstreaming air and they settle down very slowly by cutting the airflow. This particle assembly gives the bulk-mean particle volume fraction of 0.55 that is the loosest packing under no gravity.

A computer simulation experiment gives the loose packing of $\phi = 0.582$ where each particle falling from random position rolls down on existing particles until it is supported by three points to become stable against the z-axis. This is the packing under the weak gravity.

In summary, the bulk-mean particle volume fraction becomes $\phi = 0.55$ for the random loosest packing and $\phi = 0.6366 \fallingdotseq 0.64$ for the random tightest packing. And there exists no random packing in the range of $\phi > 0.64$, where only two regular arrangements are possible to exist at

$$\phi = 2\pi/9 = 0.698 \quad \text{and}$$

$$\phi = \pi\sqrt{2}/6 = 0.740$$

In order to make tighter packing than 0.64, therefore, the uniform compaction to the above regular arrangement is impossible in general so that the local aggregation is only possible to occur. This may explain the reason for the multicrystallizing phenomena of various substances.

The pores exist in the random packing and their size distribution function was derived in Section 4.6.2.

In the random packing, the number of contacting particles on a particle under consideration ranges $n = 2-10$ for $\phi = 0.64$ and $n = 3-9$ for $\phi = 0.58$. The average becomes $\langle n \rangle = 6.0$ for the both cases. The range of n is not important and found changeable by the accuracy of "the contact."

A particle becomes stable against a direction by the support of three points so that taking the other direction into account, it is supported by six particles in the random packing. The average contacting number N is called the coordination number. $N = 10$ for $\phi = 0.698$ and $N = 12$ for $\phi = 0.74$.

The subject of contact and noncontact is the same as that of integer and real number, and it is beyond computer accuracy.

7.2 Random Dispersion

Next, consider the random dispersion of equal spheres. In the region of $\phi < 0.55$, each particle has no contact point and is free from others. However, there exists adhesive force between particles in practice, giving rise to the agglomeration.

Here the agglomeration radius R is defined in unit of the particle diameter, and particles within the radius R are supposed to be connected with each other. $R = 1.0$ expresses the contacting state of particles. The size distribution of agglomerates inherent in the random dispersion was discussed in Section 4.6.3.

In the dilute case of $\phi < 0.235$, each particle is separated from others, but in the case of $0.235 < \phi < 0.55$, the agglomeration occurs easily and its size distribution becomes as explained in Section 4.6.3. The critical state of $\phi = 0.235$ will be explained later.

In summary, each particle exists separately from others in the range of $\phi = 0-0.235$, agglomerates appear in the range of $\phi = 0.235-0.55$, the particle assembly becomes the random packing in the range of $\phi = 0.55-0.64$, and the multicrystallization, i.e., the local regularity occurs in the range of $\phi > 0.64$.

7.3   Molecular System and Particle Assemblies

The morphological view of particle assemblies in the previous section can be adopted to molecular systems: The liquid state at the solidification temperature is comparable to the tightest packing of molecules and the coordination number is compared in good agreement with the experiments for various substances by X-ray diffraction analysis (Gotoh, 1971b).

On the other hand, liquids evaporate at boiling temperature. This state is equivalent to the loosest packing of molecules.

When $\phi = 0.235$, the molecular system becomes the critical state, where the liquid and the gas become indistinguishable with each other. In the range of $\phi = 0.235-0.55$, the molecular system becomes gas-like liquid.

In this fashion, the morphological comparison is made for molecular system and random particle assemblies as shown in Figure 7.1.

So far we have dealt with the spherical particles only. For other cases, an ellipsoid is considered as the model of nonspherical particles, where the shape parameter λ is devised from the ellipsoidal eccentricity. The morphological study on the random packing can apply to nonspherical particles by using $\lambda\phi$ in place of ϕ. $\lambda = 1.0$ for spheres, $\lambda = 1.5$ for water molecules, and $\lambda = 1.6$ for ethanol molecules (Gotoh, 1972).

The Lennard-Jones (6,12) pair potential energy between two molecules apart distance r is expressed by

$$u(r) = 4\varepsilon\{(\sigma/r)^{12} - (\sigma/r)^{6}\}$$

where

$$d = 2^{1/6}\sigma : \quad \text{molecular diameter}$$

$\phi = (\pi/6)d^3(\rho/m)$:   volume fraction of molecules

ρ is the density and m is the molecular mass. Using the pair potential $u(r)$, the equation of state for gases can be derived as follows:

$$pV/(NkT) = 1/\{1 - \lambda\phi/(\pi\sqrt{2}/6)\} - 18\phi/(\pi kT/\varepsilon)$$

This is the same type of the van der Waals equation of state. p is the pressure, V is the volume, T is the absolute temperature, N is the number of molecules, and k is the Boltzmann constant. The equation of state gives the following values at the critical point:

* $\lambda kT/\varepsilon = 8\sqrt{2}/9 = 1.26$ (experiment : 1.28)

* $(\pi\sqrt{2}/6)/(\lambda\phi) = 3$ (experiment : 3.15)

     $\therefore$ $\lambda\phi = (\pi\sqrt{2}/6)/3.15 = 0.235$ by use of the experiment

Accordingly, we can obtain ε and σ from experiments of the critical temperature and the critical volume, where $\lambda\phi = 0.235$ is recommended. In the case of spherical molecules, $\lambda = 1.0$ and hence, $\phi = 0.235$ is the critical state. In the range of $\phi = 0-0.235$, molecules are separated from each other; while in the range of $\phi > 0.235$, they are in the state of gas-like liquid by making agglomerates of molecules.

Let us consider the mixing of ethanol ($\lambda = 1.6$) and water ($\lambda = 1.5$) as an example of the morphological study. Both substances are in the liquid state of the molecular volume fraction $\phi \fallingdotseq 0.4$ at ordinary temperatures so that 60% of the space is empty, into which other kinds of molecules are likely to intrude. The molecular size of ethanol is 1.4 times larger than water molecule.

Water molecules are very sticky with each other, while ethanol molecules are not. The difference is easy to understand from the boiling temperatures: 78.3°C for ethanol and 100°C for water.

Ethanol molecules are likely to disperse separately into the assemblies of water molecules, but water molecules do not separate from each other and remain agglomerated, even by mixing mechanically.

References

Finney, J.L., Random packings and the structure of simple liquids. I. The geometry of random close packing, Proc. Roy. Soc., vol. A319, 479 (1970).

Gotoh, K., Molecular configuration of spherical, non-polar, dense gases and liquids, Nature Phys. Sci., vol. 232, 64–65 (1971a).

Gotoh, K., Liquid structure and the coordination of equal spheres in random assemblage, Nature Phys. Sci., vol. 231, 108 (1971b).

Gotoh, K., Liquid structure and Lennard-Jones (6,12) pair potential parameters, Nature Phys. Sci., vol. 239, 154 (1972).

Gotoh, K., Liquid and powder in view of random packing structure. Advanced Particulate Morphology, (ed. by J.K. Beddow, and T.P. Meloy), pp 171–183. CRC Press, Boca Raton, FL (1980).

Closing Remarks

The author took part in the publication of the first edition of *Powder Technology Handbook* (by K. Iinoya, K. Gotoh, and K. Higashitani, Marcel Dekker, New York, 1990). The only similar book available in those days was entitled *Micromeritics*. The revised edition of the handbook was published in 1997 as the textbook for students and researchers, and our professor, Dr. Koichi Iinoya, was celebrated on the occasion of his 80th birthday. Since then, the third edition of the powder technology handbook has been published by H. Masuda, K. Higashitani, and H. Yoshida (Taylor & Francis, Boca Raton, FL, 2006).

Powder technology covers various areas of science and technology of powder/particulate materials. It is therefore too wide and transversal for us to make a single fundamental basis such as fluid mechanics and thermodynamics. Moreover, new areas have been growing rapidly concerning environment, biotechnology, nanotechnology, computer simulation, etc. Hence, a single scholarly basis might be impossible to cover that many areas.

Accordingly, the mathematical procedures used in the author's own research subjects are compiled in this text as an example of fundamental scholarly basis, asking for further advancement in the field of engineering sciences.

Printed in the United States
By Bookmasters